*Practical methods in*
*electron microscopy*
*Volume 3*

# Practical Methods in
# ELECTRON MICROSCOPY

Volume 3

*Edited by*
## AUDREY M. GLAUERT
*Strangeways Research Laboratory*
*Cambridge*

1974
NORTH-HOLLAND PUBLISHING COMPANY – AMSTERDAM · OXFORD
AMERICAN ELSEVIER PUBLISHING CO., INC. – NEW YORK

# Part I

# FIXATION, DEHYDRATION
# AND EMBEDDING OF
# BIOLOGICAL SPECIMENS

*Audrey M. Glauert*

---

# Part II

# ULTRAMICROTOMY

*Norma Reid*

1974

NORTH-HOLLAND PUBLISHING COMPANY – AMSTERDAM · OXFORD

AMERICAN ELSEVIER PUBLISHING CO., INC. – NEW YORK

© 1974 North-Holland Publishing Company – Amsterdam

No part of this book may be reproduced in any form by print, photoprint, microfilm or any other means without written permission from the publishers

Library of Congress Catalog Card Number 74–76062
North-Holland ISBN for the series 0 7204 4250 8
North-Holland ISBN for this volume 0 7204 4256 7
American Elsevier ISBN 0 444 106650

Publishers:

NORTH-HOLLAND PUBLISHING COMPANY – AMSTERDAM
NORTH-HOLLAND PUBLISHING COMPANY, LTD. – OXFORD

Sole Distributors for the U.S.A. and Canada:

AMERICAN ELSEVIER PUBLISHING COMPANY, INC.
52 VANDERBILT AVENUE
NEW YORK, N.Y. 10017

Printed in The Netherlands

# Editor's preface

Electron microscopy is now a standard technique with wide applications in all branches of Science and Technology, and every year a large number of students and research workers start to use the electron microscope and require to be introduced to the instrument and to the techniques for the preparation of specimens. Many books are available describing the techniques of electron microscopy in general terms, but the authors of Practical Methods in Electron Microscopy consider that there is an urgent need for a comprehensive series of laboratory handbooks in which all the techniques of electron microscopy are described in sufficient detail to enable the isolated worker to carry them out successfully. The series of books will eventually cover the whole range of techniques for electron microscopy, including the instrument itself, methods of specimen preparation in biology and the materials sciences, and the analysis of electron micrographs. Only well-established techniques which have been used successfully outside their laboratory of origin will be included.

Great care has been taken in the selection of the authors since it is well known that it is not possible to describe a technique with sufficient practical detail for it to be followed accurately unless one is familiar with the technique oneself. This fact is only too obvious in certain 'one author' texts in which the information provided quickly ceases to be of any practical value once the author moves outside the field of his own experience.

Each book of the series will start from first principles, assuming no specialist knowledge, and will be complete in itself. Following the successful innovation, made by the same publishers in the parallel series Laboratory

v

Techniques in Biochemistry and Molecular Biology (edited by T. S. Work and E. Work), each book will be included, together with one or two others of the series, in a hardback edition suitable for libraries and will also be available in a cheap edition for individual use in the laboratory. Each book will be revised, independently of the others, at such times as the authors and editor consider necessary, thus keeping the series of books continuously up-to-date.

*Strangeways Research Laboratory*                    AUDREY M. GLAUERT, Sc. D.
*Cambridge, England*                                 *General editor*

# Contents of parts I and II

## PART I

### FIXATION, DEHYDRATION AND EMBEDDING OF BIOLOGICAL SPECIMENS

by Audrey M. Glauert

## PART II

### ULTRAMICROTOMY

by Norma Reid

# FIXATION, DEHYDRATION AND EMBEDDING OF BIOLOGICAL SPECIMENS

AUDREY M. GLAUERT

*Strangeways Research Laboratory*

*Cambridge, England*

# Acknowledgements

I am extremely grateful for the valuable help and constructive criticism I received from Dr. J. R. Gibbins, Dr. P. R. Lewis, Dr. G. Millonig, Mr. R. A. Parker and Mrs. M. E. Van Steveninck during the preparation of this book. In particular I wish to thank Dr. Lewis for contributing the section on perfusion fixation, and Mrs. Van Steveninck for the section on the fixation of botanical specimens in Chapter 3.

I am also indebted to the following for permission to reproduce micrographs and figures: Dr. T. C. Appleton, Dr. J. G. Bluemink, Dr. M. W. Brightman, Dr. S. Bullivant, Dr. S. Busson-Mabillot, Dr. F. Carson, Dr. E. Dimmock, Dr. C. Franzini-Armstrong, Dr. J. Gil, Dr. P. V. Johnston, Dr. H. Kushida, Dr. S. Malamed, Professor R. G. Monroe, Dr. B. I. Roots, Dr. J. McD. Tormey and Professor E. R. Weibel.

Finally I would like to thank Mr. R. A. Parker for assistance with the illustrations and for providing Figures 3.2, 5.4, 5.5 and 5.6, Mrs. Doreen Stebbings for typing the manuscript and my husband for his constant encouragement and patience.

*November 1973*                                                                 AUDREY M. GLAUERT

# Contents of Part I

Chapter 1

# Introduction

One of the major consequences of the development of techniques for the fixation, embedding and sectioning of specimens for electron microscopy has been a complete reappraisal of the micro-anatomy of biological tissues and organisms during the past 25 years. The fact that electron micrographs of ultrathin sections are used to illustrate practically every textbook and monograph, and in many research papers, in cell biology and anatomy is sufficient indication of the importance of the technique as one of the basic methods of modern biology and medical science.

This book describes in detail the methods for the fixation, dehydration and embedding of biological specimens, while techniques for preparing ultrathin sections (Reid 1974) and for staining sections to provide contrast in the electron microscope (Lewis et al. 1974) are described in companion volumes. In common with other books in this series the aim has been to describe the techniques of electron microscopy in sufficient detail to enable the isolated worker to carry them out successfully. No specialist knowledge is assumed. Only well-established techniques are discussed in full, although many others are outlined briefly and reference given to the original papers in which they are described. The literature survey for this book ended in January 1973.

## 1.1 Electron microscopy of biological specimens; the need for ultrathin sections

It is necessary to prepare ultrathin sections of biological specimens because of the limited penetration of the electron beam in the electron microscope operating at conventional accelerating voltages up to 100 kV. For adequate penetration and resolution the specimen should not be thicker than 100 nm, and preferably considerably thinner. Only a few specimens, such as virus particles and isolated sub-cellular organelles, are thin enough to be viewed directly, and consequently considerable attention has been paid to the development of fixation, embedding and sectioning techniques.

Ultrathin sections are not suitable for the study of fine structure at the optimum resolution of the electron microscope (0.5 nm for most standard instruments) since structural detail smaller than about 2 nm does not appear to be preserved by the fixation and embedding procedures at present available. Fortunately biological specimens possess a great wealth of structural detail in the 2 to 100 nm range and here the study of sections provides information that can be obtained in no other way.

There is a well-defined series of steps in the fixation, dehydration and embedding of biological specimens and in general it is only necessary to adopt a very standard routine which is now followed in many laboratories. Each of the steps of the preparative procedure is described in detail in the following chapters, together with the modifications of the basic methods that are required for special types of specimen.

## 1.2   The aim of the procedure

The final aim of the whole procedure is to produce blocks which can be sectioned without undue difficulty and which contain specimens in which the fine structure is preserved with as little alteration as possible when compared with the living organism. As shown by Reid (1974) the production of sufficiently thin sections is usually no problem with standard embedding media, but the problem of how to preserve the specimen in a life-like state is by no means solved. At each stage of the procedure, from the initial fixation onwards, changes are inevitably introduced; material is extracted, dimensions are altered (Fig. 1.1) and molecular rearrangement occurs. The best that can be done is to keep these changes to a minimum and the achievement of this is the over-riding consideration in the choice of preparative procedures.

## 1.3   Criteria of good preservation

Since some structures observed in electron micrographs of ultrathin sections cannot be visualised by any other method, there is an inevitable element of subjectivity in judging the quality of preservation achieved. There are a number of criteria, however, which can be applied:

(1) The final appearance of the specimen should be similar to that of

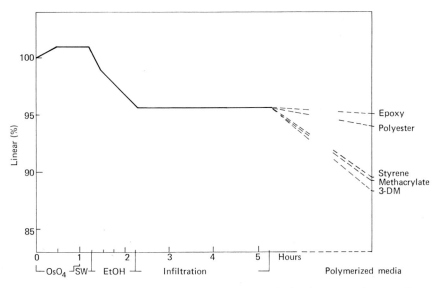

Fig. 1.1. An example of dimensional changes occurring during the preparative procedure. The graph shows the percentage changes in the diameters of sea urchin eggs during fixation in osmium tetroxide ($OsO_4$), washing in sea water (SW), dehydration in ethanol (EtOH), and infiltration and embedding in five different embedding media. 3-DM is three-dimensional methacrylate (from Kushida 1962).

the living specimen viewed with the light microscope. It is rarely profitable to examine a specimen in the electron microscope before a full understanding of its appearance in the light microscope has been obtained. The examination of 'thick' (0.5 to 2.0 μm) sections of blocks prepared for electron microscopy (Reid 1974) has recently become a powerful tool in cytology.

(2) Obvious structural damage should be absent. Membranes should be continuous; the space between the two nuclear membranes (the peri-nuclear space) should not be swollen and empty; there should be no empty spaces in the ground substance or within organelles.

(3) Loss of material and dimensional changes should be kept to a minimum. If the specimen contains a regular periodic structure, such as a myofibril, this can be used as an internal standard to follow dimensional changes.

(4) A structure should have the same general appearance after fixation with a number of different fixatives.

Different specimens require different preparative procedures, particularly at the fixation stage (Chapter 3) and the beginner will find that it is always worthwhile to first follow methods already used successfully by others for similar specimens, and then to introduce modifications if these prove to be necessary.

REFERENCES

Kushida, H. (1962), A study of cellular swelling and shrinkage during fixation, dehydration and embedding in various standard media, J. Electron Microscopy *11*, 135.
Lewis, P. R., D. P. Knight and M. A. Williams (1974), Staining methods for thin sections, in: Practical methods in electron microscopy, A. M. Glauert, ed. (North-Holland, Amsterdam).
Reid, N. (1974), Ultramicrotomy, in: Practical methods in electron microscopy, A. M. Glauert, ed. (North-Holland, Amsterdam).

# Chapter 2

# Fixatives

Fixation is the first step in the preparation of biological specimens for examination by electron microscopy of thin sections and has the aim of 'stabilising cellular organisation to such an extent that ultrastructural relations are preserved despite the subsequent rather drastic treatments of dehydration, embedding and exposure to the electron beam' (Riemersma 1968). Inevitably, however, the structure of the living specimen is modified during this stabilisation procedure, and these modifications may well be greatest for those fixatives that produce the best protection against the effects of the subsequent steps in the preparative procedure. With some fixatives obvious disruption of fine structure is observed; discontinuities are present in membranes, tubules are transformed into chains of vesicles and, with certain fixatives such as permanganates, cell components (ribosomes, myofilaments, cytoplasmic fibrils) are completely destroyed. Some of these effects depend as much on the nature of the specimen as on the fixative since they result from interactions between the two, and in consequence a number of conflicting criteria influence the choice of a fixative for a particular specimen.

In extreme examples the fixative may induce noticeable physiological changes, such as the contraction or relaxation of muscle fibres or the triggering of secretion by solutions containing calcium. The effects of individual fixatives are described in the separate sections below, but it is important to realise from the outset that there is no one perfect fixative for all types of specimen; not only do different tissues require different fixatives, but it may even be necessary to use different fixatives at different stages of development of a tissue (e.g. Kelly and Zacks 1969) or for different structural elements within a tissue (e.g. Doggenweiler and Heuser 1967). The general aim is usually to obtain satisfactory preservation of the tissue or organ as a whole, and not to achieve the best possible preservation of a given cell type with disregard to the fate of others (Palay et al. 1962). For some studies, however, it will be necessary to choose the conditions so as to obtain the optimum fixation for certain structures even if this is to the detriment of others.

The composition and properties of the various fixatives currently used in electron microscopy are described in this chapter, while suggestions on the choice of a fixative for a particular study are made in the following chapter (Chapter 3) which is concerned with methods of fixation.

## 2.1   The development of fixatives for electron microscopy

A fixative consists of a fixing agent (e.g. osmium tetroxide) in a suitable vehicle (usually a buffer plus various salts, § 2.2).

Most of the fixatives developed for light microscopy are inadequate for electron microscopy since much of the very fine detail visible by electron microscopy is not preserved, and extensive extraction and precipitation occurs (Baker 1965). Osmium tetroxide is the best of the light microscope fixatives for preserving fine structure and it was the basis of the earliest fixatives specially developed for electron microscopy (Palade 1952; Rhodin 1954; Zetterqvist 1956).

The adequate fixation of botanical tissues was less easily obtained with osmium tetroxide (Mollenhauer 1959; Afzelius 1962) and consequently potassium permanganate (Luft 1956) became popular as an alternative fixative both for plant tissues and for membranes (e.g. Robertson 1959).

A major change in fixation procedures followed the publication in 1963 by Sabatini, Bensch and Barrnett of the results of their search for fixatives suitable for electron microscope cytochemistry. They recommended the use of aldehydes, particularly glutaraldehyde, as primary fixatives, followed by a secondary fixation with osmium tetroxide. These studies made a major contribution to the development of new techniques in cytochemistry (see Lewis et al. 1974), but the importance of the work in the present context is that it drew attention to the value of aldehydes as fixatives for electron microscopy. Other aldehydes had been used previously, but not glutaraldehyde, and it was this aldehyde which was found to give excellent preservation of a wide range of animal and plant tissues when followed by a second fixation with osmium tetroxide. The procedure has the advantage that most specimens can be stored in buffer after glutaraldehyde fixation for considerable periods of time before further processing. The convenience of this double fixation and the good preservation of fine structure obtained has made it the standard procedure in most laboratories. The success of glutaraldehyde led to a re-examination of other aldehydes as primary fixatives, with the result that it was found that formaldehyde, freshly prepared from paraformaldehyde (Robertson et al. 1963) compares favourably with glutaraldehyde and has the added advantage of more rapid penetration into the specimen. Subsequently various mixtures of aldehydes have been tested and some of them, particularly a mixture of paraformaldehyde and glutaraldehyde (Karnovsky 1965), have proved to be superior to either aldehyde used alone for many specimens.

Recent years have seen the further developments of fixatives for electron microscopy, including the use of uranyl acetate as a third fixative, following glutaraldehyde and osmium tetroxide. When uranyl acetate is used in this way, before dehydration and embedding, it is often described as a 'block stain' and its properties as such are considered by Lewis et al. (1974). However, under these conditions uranyl acetate also acts as a fixative, particularly for lipid components, and so it will be included here also.

All these various fixatives and their properties are described in the sections below. In general, when starting a study of a particular specimen it is advisable to first follow the most recent methods used successfully by others (see Chapter 3) and then to make any modifications required.

## 2.2 Vehicles for fixatives

An ideal fixative would exactly match the natural environment of a living tissue with respect to pH, osmolarity and ionic constitution. For isolated cells (see § 3.6.2) an approximate match is obtained by dissolving the fixing agent in the growth medium or by exposing cells suspended in their growth medium to the vapours of the fixative, but problems arise when fixing uni- or multi-cellular organisms which live in environments of very low or very high ionic strength, or at alkaline or acid pH (§ 3.9.5), and when fixing organs and tissues. The best choice of components for the vehicle of the fixative is usually a matter of trial and error since there is little detailed knowledge of the constituents of the fluids and matrices surrounding cells or of the intracellular composition of the cells in various tissues. In addition, the fixing agent itself may have unpredictable effects on the permeability of membranes and therefore of cells and organelles (§ 2.2.3). Fortunately when a piece of tissue is immersed in a fixative the strong buffering capacity of the tissue proteins tends to keep the pH within a narrow range (Millonig and Marinozzi 1968). For this reason good fixation of some tissues can be obtained with a number of different fixatives, or even with unbuffered fixatives (Claude 1962; Malhotra 1962). Other tissues, such as the epithelia of the kidney (Rhodin 1954) and intestine (Zetterqvist 1956), and isolated cells, are much more sensitive to the composition of the fixative.

### 2.2.1   CHOICE OF VEHICLE

The properties required of a vehicle for a fixative are:

   (i)   Ability to maintain a constant pH during fixation

   (ii)   Suitable osmolarity (when mixed with the fixing agent) so that cells and organelles neither swell nor shrink during fixation

   (iii)   Suitable ionic constitution so that materials are neither extracted or precipitated during fixation.

Additional criteria, such as lack of toxicity and colloid osmotic pressure, govern the choice of vehicles for fixation by perfusion (see § 3.3.3).

The buffers commonly used for electron microscopy include phosphate, cacodylate, veronal-acetate and collidine, the majority of fixatives being buffered with phosphate or cacodylate. Veronal-acetate and collidine buffers will be included for completeness and reasons given why they have largely been superseded as buffers for primary fixatives. They are suitable as buffers for secondary fixatives but they suffer from a number of practical disadvantages and are best avoided, if possible. In fact it does not appear to be generally realised that all the buffers used in electron microscopy, with the notable exception of phosphate buffers, are potentially harmful, both to the specimen and to the microscopist, since they contain toxic components such as barbiturates (veronal–acetate) and arsenical compounds (cacodylate). They also have an unpleasant smell, which is worst for collidine. Great care should be taken not to inhale the vapours from these buffers; they should always be handled in a fume cupboard and should be flushed away with large quantities of water. It is strongly recommended that phosphate buffers be used whenever possible, particularly for routine work.

### 2.2.2   pH OF BUFFERS

Fixatives for electron microscopy are usually adjusted to a pH near neutrality using the chosen buffer system. For the majority of tissues there is little evidence that the pH of the fixative is at all critical within the range pH 6.5 to 8.0, although from model experiments with albumin, Millonig and Marinozzi (1968) suggest that phosphate buffered osmium tetroxide fixatives should be adjusted to a pH of 7.0 or below. Slightly acidic (pH 6.0) osmium tetroxide fixatives have been recommended for the preservation of nuclear material and spindle fibres. With the advent of aldehyde fixatives, primary fixation with glutaraldehyde at approximately neutral pH is recommended for all specimens.

### 2.2.3 OSMOLARITY

It is not possible to calculate the correct osmolarity for a fixative since insufficient is known about the internal osmotic pressure of living cells. In addition, it is clear that alterations in the permeability of membranes occur during fixation. After osmium tetroxide fixation membranes become freely permeable to macromolecules such as ferritin (Tormey 1965), while after glutaraldehyde or formaldehyde fixation the permeability properties of membranes are only partially preserved (Jard et al. 1966). Although there are rapid changes in the osmotic properties of cell envelopes at fixation, there is no loss of the intracellular colloid-osmotic equilibrium. Therefore fixing and rinsing solutions have to be adjusted to this equilibrium which, in terms of osmotic pressure, may differ from the iso-osmotic environment of the cell (Millonig and Marinozzi 1968).

In these circumstances the only practical course is to select a fixative following procedures used in earlier studies on a similar tissue and then to make adjustments if signs of swelling or shrinkage are observed. If a number of different cell types is present it may be necessary to use slightly different fixatives for the optimum preservation of each cell type, since there is evidence that the osmotic pressure is different in different cells and is generally higher than that of blood plasma (Palade 1956). In general a hypertonic fixative is less damaging than a hypotonic one.

It is sometimes necessary to record the osmolarity of a fixative when studying osmotically sensitive specimens so that the fixative may be standardised from one experiment to another. Workers for whom osmolarity is important, such as those studying isolated cells, favour the estimation of osmolarity from the measurement of freezing-point depression with an osmometer, since calculations from formulae do not take into account the degree of dissociation of the various salts (Millonig 1962; Maser et al. 1967).

The osmolarity of the fixative is adjusted by altering the concentration of the buffer (and not the concentration of the fixing agent) or by the addition of sodium chloride or non-ionic compounds, such as glucose or sucrose. Glucose and sucrose should not be used in primary osmium tetroxide fixatives since they inhibit proper 'fixation' of albumin in model experiments (Millonig 1964; Millonig and Marinozzi 1968), and may lead to an increased extraction of soluble material during fixation. They are suitable, however, for secondary osmium tetroxide fixatives and for aldehyde fixatives. Monovalent ions are preferred to divalent ions for the adjustment of osmolarity since divalent ions are more likely to produce specific ion effects and may induce a granular precipitation of proteins (Millonig and Marinozzi 1968). Sodium chloride is usually used and should be added to the buffer before the pH is finally adjusted since it may alter the pH. Glutaraldehyde and formaldehyde are not themselves osmotically active (§§ 2.4.1 and 2.5.1) and it has been suggested that their contribution to the osmolarity of the fixative can be ignored (Bone and Ryan 1972).

### 2.2.4  IONIC CONSTITUTION

When the composition of the fixative differs from that of the natural extracellular fluids of the tissue there is a likelihood of extraction of cellular materials and/or deposition of fixative components. These effects occur in the period before the tissue is adequately 'fixed' and are particularly marked with fixatives in which the fixing agent penetrates (e.g. osmium tetroxide) or reacts (e.g. formaldehyde) slowly. Consequently when osmium tetroxide or formaldehyde is used as the primary fixative the ionic constitution of the vehicle is important (Wood and Luft 1965; Carson et al. 1972) and care should be taken to match it as closely as possible to the tissue fluids. Also non-ionic additives should be used to adjust the osmolarity of the fixative if there is any evidence that the addition of monovalent ions will have any effect.

It seems likely that the superiority of glutaraldehyde over osmium tetroxide as a primary fixative, particularly for non-membraneous components of the cytoplasm (Figs. 2.1a and b) and extracellular materials, is partly due to the faster rate of reaction of glutaraldehyde with the result that fixation occurs before the extraction of cell components by the buffer is noticeable. This would explain why the choice of vehicle is important in the fixation of tissues with osmium tetroxide (Trump and Ericsson 1965), but not with glutaraldehyde (Gil and Weibel 1968) and why no differences are found in the results with differently buffered glutaraldehyde fixatives so long as the buffers have approximately the same osmolarity (Maunsbach 1966b; Busson-Mabillot 1971).

It is sometimes possible to relate these extractive effects directly to the presence or absence of specific ions. For example, bicarbonate ions have a specific destructive effect on some microtubules (Schultz and Case 1968)

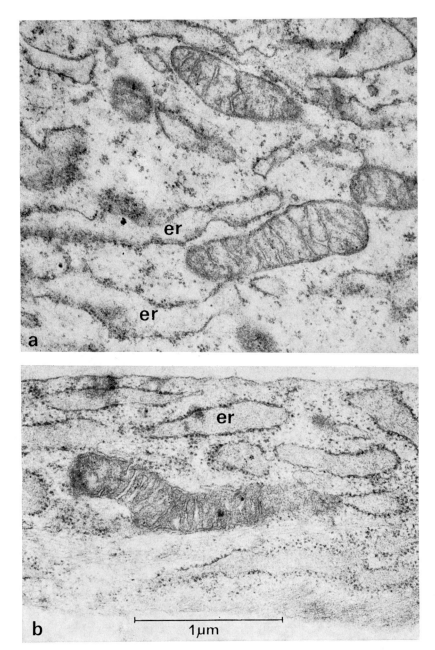

Fig. 2.1. Rat dermal fibroblasts in monolayer culture fixed (a) with osmium tetroxide and (b) with glutaraldehyde, followed by osmium tetroxide. The double fixation gives much better preservation of the contents of the cisternae of the endoplasmic reticulum (er) and of cytoplasmic filaments. (Unpublished micrographs from a study by Audrey M. Glauert and Mary R. Daniel.)

while divalent cations are necessary for the preservation of membrane lipids (e.g. Mitchell 1969). The presence of divalent cations in fixatives is not wholly beneficial since this may induce a precipitation of proteins and give a granular appearance to cell structures (Millonig and Marinozzi 1968). Magnesium is preferred to calcium by Millonig (1973) because it is a smaller molecule and, at reasonable concentrations, does not precipitate proteins. In addition calcium may exert significant physiological effects, such as the stimulation of muscle fibres or the triggering of secretion.

Artefacts due to precipitation are less easy to detect than those due to extraction since such precipitates are unlikely to be noticed unless they are comparatively large and dense. A striking and extreme example of such precipitates is provided by the observation of Gil and Weibel (1968) that spherical dense granules appear on the alveolar surfaces of rat lungs and on erythrocytes after fixation with glutaraldehyde followed by osmium tetroxide in phosphate buffer (Fig. 2.2). The appearance of the granules is independent of the buffer used with the glutaraldehyde or for washing, and the granules are only seen when the osmium tetroxide is buffered with phosphate and not with other buffers. They also do not appear after primary fixation with phosphate buffered osmium tetroxide, and Gil and Weibel (1968) concluded that the precipitates are a complex containing osmium, phosphate and an unidentified substrate from the tissue which is modified by glutaraldehyde fixation.

In general, the possibility of the formation of precipitates during fixation should be borne in mind, particularly when using fixatives containing divalent cations or phosphates.

### 2.2.5  PHOSPHATE BUFFERS

Phosphate buffers have the advantage that they are the most 'physiological' of the buffers used in electron microscopy since they mimic certain components of extracellular fluids and are non-toxic to cells in culture. In consequence they are excellent buffers for fixation by perfusion and for slowly penetrating and reacting fixatives, such as osmium tetroxide. Their only disadvantages are that precipitates are more likely to occur during fixation than with other buffers (see § 2.2.4), and that they become slowly contaminated with micro-organisms.

A number of different phosphate buffers have been used and there is little evidence that one is basically superior to another so long as the osmolarity is the same. Most of the buffers used are based on Sörensen's buffer (see Dawson et al. 1969) which is a mixture of monobasic and dibasic

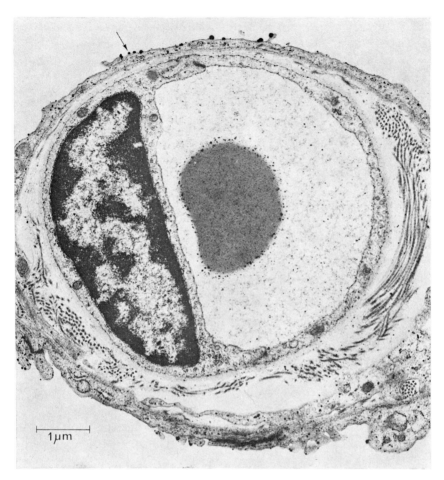

Fig. 2.2.   Dog lung fixed with phosphate buffered glutaraldehyde, washed in phosphate buffer, and post-fixed with phosphate buffered osmium tetroxide. Dense granules are present on the alveolar surface (arrow) and on the surface of an erythrocyte in a capillary. (Unpublished micrograph from a study by Joan Gil and Ewald R. Weibel.)

sodium phosphates. The method of preparation of a 0.1 M phosphate buffer by the method of Sörensen is therefore described first, and then the formulae of the buffers that have been widely used in electron microscopy are given. A phosphate buffer of any desired pH and osmolarity can be prepared following the same basic principles. The pH of phosphate buffers changes little with temperature; the buffers are non-toxic and are stable for several weeks at 4 °C if sucrose or glucose have not been added. Sediments may appear during storage and so it is advisable to make the buffers up freshly each month from the stock solutions.

### Method of preparation of phosphate buffer I

(Sörensen, see Dawson et al. 1969)

1.  Prepare a 0.2 M solution of dibasic sodium phosphate with

    |  |  |  |  |
    |---|---|---|---|
    |  | $Na_2HPO_4.$ | $2H_2O$ | 35.61 g |
    | or | $Na_2HPO_4.$ | $7H_2O$ | 53.65 g |
    | or | $Na_2HPO_4.$ | $12H_2O$ | 71.64 g |
    |  | distilled water to make | 1000 | ml |

2.  Prepare a 0.2 M solution of monobasic sodium phosphate with

    |  |  |  |  |
    |---|---|---|---|
    |  | $NaH_2PO_4.$ | $H_2O$ | 27.6 g |
    | or | $NaH_2PO_4.$ | $2H_2O$ | 31.21 g |
    |  | distilled water to make | 1000 | ml |

3.  Prepare the 0.1 M phosphate buffer by mixing $x$ ml of 0.2 M dibasic sodium phosphate with $y$ ml of 0.2 M monobasic sodium phosphate and diluting to 100 ml with distilled water.

    | pH (at 25°C) | $x$ ml | $y$ ml |
    |---|---|---|
    | 5.8 | 4.0 | 46.0 |
    | 6.0 | 6.15 | 43.85 |
    | 6.2 | 9.25 | 40.75 |
    | 6.4 | 13.25 | 36.75 |
    | 6.6 | 18.75 | 31.25 |
    | 6.8 | 24.5 | 25.5 |
    | 7.0 | 30.5 | 19.5 |
    | 7.2 | 36.0 | 14.0 |
    | 7.4 | 40.5 | 9.5 |
    | 7.6 | 43.5 | 6.5 |
    | 7.8 | 45.75 | 4.25 |
    | 8.0 | 47.35 | 2.65 |

The osmolarity of the buffer is adjusted by varying the molarity of the phosphates, or by the addition of sucrose, glucose or sodium chloride. The osmolarity of 0.1 M Sörensen's buffer at pH 7.2 is 226 mosmols (Fahimi and Drochmans 1965b); corresponding values for 0.05 M, 0.075 M and 0.15 M buffers are 118, 180 and 350 mosmols, while addition of 0.18 M sucrose to a 0.1 M buffer raises the osmolarity to 425 mosmols.

## Method of preparation of phosphate buffer II
(Maunsbach 1966a)

Prepare the 0.135 M phosphate buffer with

|     | $NaH_2PO_4$. $H_2O$ | 2.98 g |
| --- | --- | --- |
| and | $Na_2HPO_4$. $7H_2O$ | 30.40 g |
|     | distilled water to make | 1000 ml |

The pH of this buffer is 7.35 and the osmolarity is 298 mosmols. Maunsbach (1966a) used this buffer for fixation of the rat kidney by perfusion with osmium tetroxide, glutaraldehyde or formaldehyde.

## Method of preparation of phosphate buffer III
(Karlsson and Schultz 1965)

Prepare the phosphate buffer with

| $NaH_2PO_4$. $H_2O$ | 3.31 g |
| --- | --- |
| $Na_2HPO_4$. $7H_2O$ | 33.77 g |
| distilled water to make | 1000 ml |

The pH of this buffer is 7.4; the osmolarity is 320 mosmols and is equal to that of the cerebro-spinal fluid of rats. Karlsson and Schultz (1965) used this buffer for fixation of the rat central nervous system by perfusion with aldehydes.

## Method of preparation of phosphate buffer IV
(Millonig 1964; Karlsson and Schultz 1965)

Prepare the phosphate buffer with

| $NaH_2PO_4$. $H_2O$ | 1.8 g |
| --- | --- |
| $Na_2HPO_4$. $7H_2O$ | 23.25 g |
| NaCl | 5.0 g |
| distilled water to make | 1000 ml |

The pH of this buffer is 7.4. The osmolarity is 440 mosmols and so the buffer is hypertonic to most body fluids. It was recommended by Millonig (1964) for the fixation of very hydrated tissues with osmium tetroxide. The formula of this buffer illustrates how the osmolarity of a phosphate buffer may be adjusted with sodium chloride without altering the molarity of the phosphates. Millonig (1964) suggests an even higher concentration (3%) of sodium chloride for the fixation of marine organisms.

## Method of preparation of phosphate buffer V
(Millonig 1961)

1. Prepare the following solutions

   2.26% $NaH_2PO_4$. $H_2O$ in water    (solution A)
   2.52% NaOH in water    (solution B)

2. Prepare the 0.13 M phosphate buffer with

   solution A    41.5 ml
   solution B     8.5 ml (for pH 7.3).

The pH of this buffer can be adjusted to any required value with solution B without changing the molarity. The buffer is stable for some weeks at 4 °C.

### 2.2.6  CACODYLATE BUFFERS

Cacodylate buffers were first proposed for electron microscopy by Sabatini et al. (1963). They have the advantages that they are easy to prepare, are stable during storage for long periods and do not support the growth of micro-organisms. In addition, they are convenient when a controlled amount of calcium is to be added to the fixative since precipitation does not occur at the low concentrations of calcium (1 to 3 mM) that are usually required. Their main disadvantages are that they contain arsenic, which is toxic and may act as a fixative, and they have an unpleasant smell. They should always be handled in a fume cupboard.

There are few records of the use of cacodylate buffers for primary fixation with osmium tetroxide. When used during primary fixation with aldehydes the quality of preservation is usually similar to that with phosphate buffers, although they give less uniformly good results in the fixation of blood and bone marrow cells with paraformaldehyde (Carson et al. 1972) (Fig. 2.3 and Table 2.1).

## Method of preparation of cacodylate buffers

1. Prepare a 0.4 M solution of sodium cacodylate with
   $Na(CH_3)_2AsO_2$. $3H_2O$    21.4 g
   distilled water to make    250    ml

2. Prepare the 0.2 M cacodylate buffer with

   0.4 M sodium cacodylate    50 ml
   0.2 M HCl                            8 ml (approx. for pH 7.2)
   distilled water to make    100 ml

The pH of the buffer is adjusted to the required value with the HCl.

TABLE 2.1

Summary of cytological effects of buffer and temperature variation on paraformaldehyde fixation. (From Carson et al. 1972.)

| | s-Collidine | Cacodylate | | Phosphate | |
| --- | --- | --- | --- | --- | --- |
| | (4°C) | (4°C) | (Room temperature) | (4°C) | (Room temperature) |
| Membranes | Disrupted | Good | Good | Good | Good |
| Nucleus | | | | | |
| Chromatin aggregation | None | Peripheral | Peripheral | Peripheral | Peripheral |
| Density | Low | Moderate | Moderate | Dense | Dense |
| Perichromatin granules | Present | Present | Present | Present | Present |
| Peri-nuclear space | Nil | Increased swelling | Nil to slight | Slight | Nil |
| Mitochondria | Disrupted | Good Some mottled irregularity of matrix | Good | Excellent | Excellent |
| Endoplasmic reticulum swelling | Disrupted | Slight | Nil | Nil | Nil |
| Cytoplasmic matrix | Extracted | Varies from dense granular to irregular mottled | Some extraction | Dense uniform | Dense uniform |
| Preservation of granules | | | | | |
| Neutrophils | Poor | Poor | Fair to good | Excellent | Excellent |
| Basophils | Poor | Poor | Poor | Poor | Poor |
| Eosinophils | Fair | Excellent | Good | Excellent | Excellent |
| Platelets | Poor | Good | Good | Excellent | Excellent |

### 2.2.7   VERONAL–ACETATE BUFFERS

The Michaelis veronal–acetate buffers were chosen for use with osmium tetroxide in the early development of fixatives for electron microscopy in New York (Palade 1952) and Stockholm (Rhodin 1954; Zetterqvist 1956) because they cover the physiological range of pH at constant ionic strength and they appeared to give slightly better results than phosphate buffers (Palade 1956). Later studies have shown that veronal–acetate is a poor buffer

Fig. 2.3a

Fig. 2.3.   Thin sections of neutrophils fixed with paraformaldehyde in various buffers: (a) phosphate buffer; (b) cacodylate buffer; (c) collidine buffer. (Unpublished micrographs from a study by Freida L. Carson.)

system in the region of pH 5.2 to 7.5 (Bennett and Luft 1959) and is useless at physiological pH. It seems likely that the early successes were due to the fact that some tissues are adequately fixed with unbuffered osmium tetroxide (Claude 1962; Malhotra 1962) so that the absence of an effective buffer system was not disastrous. Veronal–acetate buffers are unsuitable for aldehyde fixatives since sodium veronal reacts with aldehydes to produce a substance with no buffering capacity in the physiological range of pH values (Holt and Hicks 1961).

The formulae for the veronal–acetate buffers used in electron microscopy are included here, partly for historical reasons, but also because they are satisfactory, but not recommended, for buffering osmium tetroxide solutions which are to be used as a second fixative, following primary fixation with an aldehyde, and they are also useful for fixation (or block staining) with uranyl acetate since, unlike phosphate and cacodylate buffers, they do not cause the formation of precipitates.

Veronal–acetate buffers contain a barbiturate and so their use should be discouraged except when no alternative is available. They should be handled and disposed of with great care. They cannot be stored for any length of time in the absence of the fixing agent since they support the growth of micro-organisms.

### *Method of preparation of veronal-acetate buffer I*

(Palade 1952)

1.  Prepare a veronal–acetate stock solution with

| | | |
|---|---|---|
| | sodium veronal (barbitone sodium) | 2.89 g |
| | sodium acetate (anhydrous) | 1.15 g |
| *or* | sodium acetate (hydrated) | 1.90 g |
| | distilled water to make | 100 ml |

This solution is stable and keeps for some months at 4 °C.

2.  Prepare the veronal–acetate buffer with

| | | |
|---|---|---|
| veronal–acetate stock solution | 5.0 ml | |
| distilled water | 15.0 ml | |
| 0.1 N HCl | 5.0 ml | (approx.) |

The HCl is added gradually until the required pH is reached. This buffer supports the growth of moulds and bacteria and consequently it cannot be stored for any length of time, even at 4 °C.

Fig. 2.3b

## Method of preparation of veronal-acetate buffer II

(Zetterqvist 1956)

1.  Prepare a veronal–acetate stock solution with

|  |  |  |
|---|---|---|
| sodium veronal (barbitone sodium) | 2.94 g | |
| sodium acetate (hydrated) | 1.94 g | |
| distilled water to make | 100 | ml |

This solution is stable and keeps for some months at 4 °C.

2. Prepare a Ringer's solution with

| | |
|---|---|
| sodium chloride | 8.05 g |
| potassium chloride | 0.42 g |
| calcium chloride | 0.18 g |
| distilled water to make | 100 ml |

3. Prepare the veronal-acetate buffer with

| | |
|---|---|
| veronal–acetate stock solution | 10.0 ml |
| Ringer's solution | 3.4 ml |
| distilled water | 25.0 ml |
| 0.1 HCl | 11.0 ml (approx.) |

The pH is adjusted to the required value with the HCl. This buffer is not stable and cannot be stored for any length of time.

### Method of preparation of veronal-acetate buffer III

(Ryter and Kellenberger 1958)

1. Prepare a veronal–acetate stock solution with

| | |
|---|---|
| sodium veronal (barbitone sodium) | 2.94 g |
| sodium acetate (hydrated) | 1.94 g |
| sodium chloride | 3.40 g |
| distilled water to make | 100 ml |

2. Prepare the veronal–acetate buffer with

| | |
|---|---|
| veronal–acetate stock solution | 5.0 ml |
| distilled water | 13.0 ml |
| 1.0 M calcium chloride | 0.25 ml |
| 0.1 N HCl | 7.0 ml (approx.) |

The pH is adjusted to the required value with the HCl. This buffer must be prepared on the day it is to be used since it quickly becomes contaminated with micro-organisms.

#### 2.2.8 COLLIDINE BUFFERS

Bennett and Luft (1959) suggested the use of *s*-collidine ($\gamma$-collidine or 2,4,6-trimethyl-pyridine) as a buffer system for electron microscopy since it exerts its maximum buffering capacity in the neighbourhood of pH 7.4 when half neutralised with a strong acid. In addition the buffer is stable indefinitely at room temperature. Collidine buffers are not suitable, however, for buffering primary osmium tetroxide fixatives since considerable extraction of tissue components occurs during fixation and the subsequent dehydration

(Wood and Luft 1965). They also lead to lysis of the cytoplasmic matrix and extensive destruction of membranes when used to buffer paraformaldehyde fixatives (Carson et al. 1972) (Fig. 2.3c and Table 2.1) and give poorer results for some tissues than phosphate or cacodylate buffers in glutaraldehyde fixatives (Busson-Mabillot 1971) (Fig. 2.8). Collidine buffers are toxic, have a particularly horrid smell and must be handled with extreme care. In

Fig. 2.3c

consequence they cannot be recommended for general use; they are included here because they appear to be adequate for secondary fixation with osmium tetroxide, following an aldehyde (e.g. Gil and Weibel 1968) and are of value for the fixation of very large blocks of tissues where their extractive effects aid the penetration of the fixative (Winborn and Seelig 1970).

It is necessary to use purified *s*-collidine. This is obtainable commercially (see Appendix) although it is expensive. Alternatively it may be purified in the laboratory by distillation (Holt and Hicks 1961).

## *Method of preparation of collidine buffers*

(Bennett and Luft 1959)

1. Prepare a stock solution of *s*-collidine with

    | | | |
    |---|---|---|
    | *s*-collidine (pure) | 2.67 | ml |
    | distilled water to make | 50 | ml (approx.) |

2. Prepare the 0.2 M collidine buffer with

    | | | |
    |---|---|---|
    | *s*-collidine stock solution | 50 | ml |
    | 1.0 N HCl | 9 | ml (approx. for pH 7.4) |
    | distilled water to make | 100 | ml |

The pH of the buffer is adjusted with the HCl. It can be stored indefinitely at room temperature.

## 2.3   Osmium tetroxide fixatives

The value of osmium tetroxide for the preservation of fine detail was apparent as long ago as 1927 when Strangeways and Canti studied the effects of various fixatives on the appearance of living cells by dark-field microscopy. They found that osmium tetroxide was the only fixative of those tested that preserved delicate cytoplasmic processes and caused no alteration in mitochondria and fat droplets. When the first studies on ultra-thin sections were made in the late 1940s and early 1950s it was soon found that osmium tetroxide was preferable for the examination of cellular fine structure to other fixatives developed for light microscopy.

The main disadvantage of osmium tetroxide is that it penetrates and reacts with tissues so slowly that considerable changes in structure can occur before fixation is complete. The ionic constitution of the buffer (see § 2.2.4) has to be chosen carefully since it exerts a profound effect on the fine structure of the specimen (Trump and Ericsson 1965). Also the buffer influences the rate of penetration of the osmium tetroxide. Specimens for fixation must be very small (0.5 to 1.0 mm thick) and even then a central region may still be poorly preserved. In consequence osmium tetroxide has been completely superseded by aldehydes as a primary fixative and the appearance of specimens fixed with osmium tetroxide alone will not be considered in any detail. It still plays an essential role, however, as a second fixative since it reacts with components of tissues, particularly lipids, that are not fixed by aldehydes, and also acts as a stain (Millonig and Marinozzi 1968). During this second fixation the slow rate of penetration of osmium tetroxide is no longer such a disadvantage since the structure of the specimen has already been partially stabilised by the primary aldehyde fixation. Consequently the composition of the buffer is much less important in secondary osmium tetroxide fixatives than in primary fixatives.

### 2.3.1    CHEMICAL AND MORPHOLOGICAL EFFECTS OF OSMIUM TETROXIDE FIXATIVES

A considerable amount is known about the chemistry of the interactions of osmium tetroxide with isolated lipids, proteins, nucleic acids and carbohydrates and this knowledge is well summarised in reviews by Millonig and Marinozzi (1968), Riemersma (1970) and Hayat (1970). Although the results of such studies are of considerable interest to the molecular biologist, and may well lead to the development of improved fixatives in the future, as yet they are of little practical help to the microscopist in the choice of a fixative, since there remains considerable uncertainty about the way osmium tetroxide interacts with cellular components *in situ* (Riemersma 1970).

Studies of the effects produced in tissues during treatment with solutions of osmium tetroxide are of much more practical use in providing guide lines for the choice of fixatives and fixation methods, although comparative studies have shown considerable variations between one type of specimen and another so that it is difficult to make useful generalisations. In addition, few studies have yet been made of the effects of osmium tetroxide on aldehyde-fixed tissues.

An excellent summary of the results of studies of lipid extraction from tissues is given in the review by Stein and Stein (1971). The total loss of lipid from tissues fixed only with osmium tetroxide can be very high, occurring mainly during dehydration, but this loss can be reduced greatly by the addition of calcium to the fixative (e.g. Strauss and Arabian 1969) and by modifications to the dehydration procedure (see Chapter 4). Although most of the lipid is lost during dehydration, a small amount is extracted by the osmium tetroxide fixative itself and this loss is accompanied by immediate changes in membranes which become freely permeable to small ions and protein molecules (Tormey 1965; Amsterdam and Schramm 1966). At the same time proteins enclosed by membranes remain soluble and are able to diffuse out. Osmium tetroxide fixation destroys the osmotic properties of membranes and consequently the osmolarity of the vehicle of the fixative is not important. Similar results are obtained with fixatives of low and high osmolarity (e.g. Bone and Ryan 1972).

The changes in membranes produced during fixation with osmium tetroxide lead to noticeable morphological effects, such as the shrinkage of nerve myelin (Moretz et al. 1969) and the swelling of isolated cells. This swelling occurs in iso-osmotic fixatives (Bahr et al. 1957; Kushida 1962; Millonig and Marinozzi 1968) and can be prevented by the addition of sodium chloride to make the fixative slightly hypertonic. Sucrose should not be used to adjust the osmolarity of the fixative since sucrose does not inhibit the uptake of water by isolated cells, and sea urchin eggs have been observed to swell even more in a sucrose-containing fixative than in a simple osmium tetroxide solution (Millonig and Marinozzi 1968).

There are many reports in the literature of the beneficial effects of adding calcium ions to osmium tetroxide fixatives for the preservation of a variety of structures, including spindle fibres, nerve myelin, membranes and phospholipids. Many of these improvements appear to be directly related to the decreased extraction of lipids resulting from fixation in the presence of calcium ions. Consequently it is often profitable to add calcium chloride, at a final concentration of 1 to 3 mM, to osmium tetroxide fixatives, although care must be taken not to induce a granular precipitation of proteins (Millonig and Marinozzi 1968) or of other components of the fixative, such as phosphates. Millonig (1973) prefers to add magnesium chloride, rather than calcium chloride, to phosphate-buffered fixatives, because precipitates only then form at high concentrations. To date, however, there is little published work to show whether magnesium has similar beneficial effects on the retention of lipids and the preservation of fine structure.

### 2.3.2 OSMIUM TETROXIDE SOLUTIONS

**Warning:** Osmium tetroxide volatilises readily at room temperature. The vapours are very harmful to the eyes, nose and throat. Osmium tetroxide must only be handled in a fume cupboard with good ventilation.

Osmium tetroxide is supplied in the form of yellow crystals in glass ampoules, usually in amounts of 0.1 g or 1.0 g, or as an aqueous solution. Some suppliers use ampoules with pre-scored necks so that a cutting tool is not required. For some fixatives, such as those buffered with phosphates, it is customary to dissolve sufficient solid osmium tetroxide in the buffer to obtain the desired concentration (usually 1%). For other fixatives an aqueous solution (usually 2%) is required and if it is not available commercially it is prepared in the following way:

### *Method of preparation of osmium tetroxide solutions*

To prepare a 2% solution, the label on an ampoule containing 0.1 g or 1.0 g of osmium tetroxide is removed and the ampoule is broken and dropped into a brown glass bottle with a glass stopper. 5 or 50 ml of glass distilled water is then added and the bottle shaken. The bottle must be clean or the solution will quickly become discoloured.

The solution should be prepared at least one day before it is required since osmium tetroxide takes a long time to dissolve, unless a sonicator is available; then the solution can be prepared in a few minutes. It is stable for some months and is best stored on a shelf in a fume cupboard away from direct sunlight. It should not be kept in a refrigerator or other closed space since any leakage of vapour will discolour all the internal surfaces and may well affect the other contents. Osmium tetroxide solutions and vapours can penetrate plastics and so glass containers should be used for storage.

Osmium tetroxide is very expensive and some suppliers (see Appendix) are prepared to pay for used fixative which is then regenerated. Alternatively osmium tetroxide can be reclaimed in the laboratory by the oxidative-distillation procedure described by Jacobs and Liggett (1971). These procedures have the advantage of solving the disposal problems and help to conserve a rare material. If it is necessary to dispose of osmium tetroxide down a sink, very large quantities of running water should be used.

### 2.3.3 PHOSPHATE BUFFERED OSMIUM TETROXIDE FIXATIVES

Phosphate buffers (§ 2.2.5) are superior to other buffers for primary fixation with osmium tetroxide since they are better able to maintain 'physiological' conditions during the period when the osmium is slowly penetrating into and reacting with the tissues (Millonig 1961). Their only disadvantage is the possibility of granular precipitates occurring when calcium (but not magnesium) is added to the fixative (§ 2.2.4) or when phosphate buffered osmium tetroxide is used as a secondary fixative after fixation with glutaraldehyde (Gil and Weibel 1968; Fig. 2.2). However, such precipitates are rarely observed and phosphate buffers are strongly recommended for general use in both primary and secondary fixatives.

### *Method of preparation of phosphate buffered osmium tetroxide fixatives*

1. Prepare the selected phosphate buffer (§ 2.2.5) at the required pH (usually pH 7.0 to 7.4) and molarity (usually 0.1 M).

2. If necessary, adjust the osmolarity of the buffer with sodium chloride (see § 2.2.5, buffer IV) for primary fixatives, or with sucrose for secondary fixatives.

3. If required, add calcium chloride or magnesium chloride (final concentration of 1 to 3 mM) taking care to avoid the formation of a precipitate.

   (*Note:* Tests for precipitation must be done at the temperature at which the fixative is to be used.)

4. Break an ampoule containing solid osmium tetroxide and drop it into a clean, brown, glass-stoppered bottle. Add sufficient phosphate buffer to give the required final concentration of osmium tetroxide (usually 1%).

The osmium tetroxide takes some time to dissolve and so the fixative should be prepared at least a day before it is required or a sonicator should be used. The pH of the buffer does not change on the addition of the osmium tetroxide. The fixative is stable for some weeks at room temperature and is best stored in a second air-tight container on a shelf in a fume cupboard away from direct sunlight, and not in a refrigerator.

#### 2.3.4 CACODYLATE BUFFERED OSMIUM TETROXIDE FIXATIVES

It is not customary to use cacodylate buffers (§ 2.2.6) for osmium tetroxide fixatives, but when the primary fixation has been with a cacodylate buffered aldehyde fixative it is convenient to use the same buffer for the subsequent fixation with osmium tetroxide, and cacodylate buffers are being increasingly used for this purpose.

### *Method of preparation of cacodylate buffered osmium tetroxide fixatives*

1. Prepare the cacodylate buffer (§ 2.2.6) at the required molarity (usually 0.1 M) and pH (usually 7.2 to 7.4).

2. If necessary, adjust the osmolarity of the buffer with sucrose.

3. If required, dissolve sufficient anhydrous calcium chloride or magnesium chloride to give a final concentration of calcium or magnesium of 1 to 3 mM.

4. Break an ampoule containing 0.1 g or 1.0 g of solid osmium tetroxide

and drop it into a clean, brown, glass-stoppered bottle. Add sufficient cacodylate buffer to give the required final concentration of osmium tetroxide (usually 1%) and shake the bottle well.

The osmium tetroxide takes some time to dissolve and so the fixative is prepared at least the day before it is required, or a sonicator should be used. The fixative should be prepared and stored in a fume cupboard where it will be stable for several weeks at room temperature.

### 2.3.5  VERONAL–ACETATE BUFFERED OSMIUM TETROXIDE FIXATIVES

Veronal–acetate buffers are not suitable for primary fixation with osmium tetroxide (see § 2.2.7) but they are satisfactory, although not recommended, for buffering solutions of osmium tetroxide for use as a second fixative where the control of pH is less important.

#### *Palade's veronal-acetate buffered osmium tetroxide fixative*

This fixative was developed by Palade (1952) on the basis of comparative studies on various tissues of rats.

#### *Method of preparation of Palade's fixative*
(Palade 1952).

1.  Prepare a veronal-acetate stock solution with

|  | | |
|---|---|---|
| | sodium veronal (barbitone sodium) | 2.89 g |
| | sodium acetate (anhydrous) | 1.15 g |
| *or* | sodium acetate (hydrated) | 1.90 g |
| | distilled water to make | 100    ml |

This solution is stable and keeps for some months at 4 °C.

2.  Prepare the veronal-acetate buffered fixative with

| | |
|---|---|
| 2% osmium tetroxide in water | 12.5 ml |
| (see § 2.3.2.) | |
| veronal–acetate stock solution | 5.0 ml |
| 0.1 N HCl | 5.0 ml (approx.) |
| distilled water to make | 25.0 ml |

The osmium tetroxide solution and veronal-acetate solution are mixed together in a brown, glass-stoppered bottle in a fume cupboard, and then the HCl is added gradually until the required pH (usually 7.2 to 7.4) is reached. The final concentration of osmium tetroxide is 1%. The fixative is stable for some weeks at room temperature and should be stored in a fume cupboard.

### Caulfield's veronal-acetate buffered osmium tetroxide fixative

Palade's (1952) fixative is hypotonic to red blood cells and Caulfield (1957) suggested that the osmolarity be adjusted by the addition of sucrose; 4.5 g of sucrose is added to 100 ml of fixative for animal tissues, and 1.5 g for plant tissues.

### Zetterqvist's veronal-acetate buffered osmium tetroxide fixative

Rhodin (1954) and Zetterqvist (1956) used Michaelis' veronal–acetate buffer with added salts. Zetterqvist used a Ringer solution and consequently his fixative contains 1 mM calcium chloride, thus making it suitable for secondary fixation with osmium tetroxide for specimens which require a low concentration of calcium to be present.

### Method of preparation of Zetterqvist's fixative

(Zetterqvist 1956)

1. Prepare a veronal–acetate stock solution with

| | |
|---|---|
| sodium veronal (barbitone sodium) | 2.94 g |
| sodium acetate (hydrated) | 1.94 g |
| distilled water to make | 100 ml |

   This solution is stable and keeps for some months at 4 °C.

2. Prepare a Ringer's solution with

| | |
|---|---|
| sodium chloride | 8.05 g |
| potassium chloride | 0.42 g |
| calcium chloride | 0.18 g |
| distilled water to make | 100 ml |

3. Prepare the veronal–acetate buffered fixative with

| | |
|---|---|
| 2% osmium tetroxide in water (see § 2.3.2.) | 25.0 ml |
| veronal–acetate stock solution | 10.0 ml |
| Ringer's solution | 3.4 ml |
| 0.1 N HCl | 11.0 ml (approx.) |

The fixative is prepared in the same way as Palade's fixative (see above) and the pH is adjusted to the required value (usually pH 7.2 to 7.4) with the HCl. The final concentration of osmium tetroxide is 1%. The fixative is stable for some weeks at room temperature and should be stored in a fume cupboard.

### *Ryter-Kellenberger veronal-acetate buffered osmium tetroxide fixative*

Bacterial nuclei are not well preserved by Palade's fixative and after a systematic study Ryter and Kellenberger (1958) suggested the use of a buffer adjusted to pH 6.0 and containing 10 mM calcium chloride. The fixative is prepared by dissolving 0.1 g of solid osmium tetroxide in veronal–acetate buffer III (see § 2.2.7). The 'standard' fixation for bacteria also includes treatment with a solution of uranyl acetate and is described in detail in § 3.9.3.

### 2.3.6   COLLIDINE BUFFERED OSMIUM TETROXIDE FIXATIVES

Collidine buffered osmium tetroxide fixatives are not suitable as primary fixatives (see § 2.2.8), and $s$-collidine is a most unpleasant substance, but the method of preparation of the fixative is included here for completeness, since it is used by some workers as a secondary fixative.

### *Method of preparation of collidine buffered osmium tetroxide fixatives*

(Bennett and Luft 1959)

1.   Prepare the collidine buffer (see § 2.2.8).

2.   Prepare the collidine buffered osmium tetroxide fixative with

> 2% osmium in water (see § 2.3.2)     2 ml
> collidine buffer                              1 ml

Mix the two components in a brown, glass-stoppered bottle. The final concentration of osmium tetroxide is 1.33%; other concentrations are obtained by varying the ratio of the two solutions. The fixative is stable for many days.

## 2.4   Glutaraldehyde fixatives

Formaldehyde has long been one of the standard fixatives for light micro-scopy and was used in some early electron microscope studies, while acrolein was investigated as a possible fixative by Luft in 1959. The potentialities of aldehydes as fixatives for electron microscopy were not fully realised, however, until Sabatini et al. (1963) examined their suitability for use in cytochemical studies. The quality of fixation produced by glutaraldehyde, glyoxal, hydroxyadipaldehyde, crotonaldehyde, pyruvic aldehyde, acetalde-hyde, acrolein, methacrolein and formaldehyde in cacodylate- or phosphate-buffered solutions was investigated and it was found that none of these were suitable as a general fixative when used alone since the contrast obtained was low and the conventional image of membranes was not obtained. The important observation made by Sabatini et al. (1963) was that if fixation with an aldehyde, particularly glutaraldehyde, was followed by a second fixation with osmium tetroxide the results were as good as the best obtained with osmium tetroxide alone. Subsequently it has been found that this double fixation is excellent for a great variety of specimens with the result that glutaraldehyde, either alone or in combination with paraformaldehyde, is the most widely used primary fixative for electron microscopy at the present time (see Chapter 3). It is particularly useful for preserving the continuity of membrane systems (e.g. Franzini-Armstrong and Porter 1964; Tormey 1964) (Figs. 2.4 and 2.5), for labile cells, such as those in embryonic and pathological tissues, for isolated cells, and for fixation by perfusion (§ 3.3.3). In addition, it provides better preservation of the contents of the endoplasmic reticulum (Fig. 2.1) and of myofibrils (Franzini-Armstrong and Porter 1964) (Fig. 2.6).

Double fixation with glutaraldehyde followed by osmium tetroxide has many advantages over fixation with osmium tetroxide alone. The size of the block of tissue is now much less critical, since relatively large pieces, some millimeters in dimensions, can be immersed in the primary fixative and smaller blocks cut out from the well preserved surface zone for subsequent fixation in osmium tetroxide. In addition, with the exception of certain delicate tissues, specimens can be stored in buffer in the cold for periods up to several months before post-fixation with no obvious alterations in fine structure.

### 2.4.1 CHEMICAL AND MORPHOLOGICAL EFFECTS OF GLUTARALDEHYDE FIXATIVES

The excellent results obtained when glutaraldehyde is used as a primary fixative are mainly due to the facts that it reacts very rapidly with proteins and that, being a dialdehyde, it stabilises structures by cross-linking before there is any opportunity for extraction by the buffer to occur. In consequence far more of the ground substance of the cytoplasm and of the extracellular matrices is preserved than after primary fixation with osmium tetroxide (Fig. 2.1).

The reactions between glutaraldehyde and components of tissues have not yet been studied systematically; the available evidence is summarised by Millonig and Marinozzi (1968) and Hayat (1970). It is clear that glutaraldehyde is an excellent fixative for proteins, even though it causes some

Fig. 2.4a

Fig. 2.4.   Thin sections of rabbit ciliary epithelium. (a) Glutaraldehyde, followed by osmium tetroxide, fixation. The complex membrane foldings represent interdigitations of adjacent cells. p, projection from the cell in the right-hand upper corner. (b) Osmium tetroxide fixation. Many long rows of vesicles are associated with the interdigitations of adjacent cells. The configuration of the rows is very similar to that of the interdigitations and arrows indicate places where two rows are joined together to form U-shaped loops in the manner of interdigitations. These vesicles are interpreted as fixation artefacts.
(From Tormey 1964.)

Fig. 2.4b

Fig. 2.5. Transverse sections of fish muscle fibres fixed with glutaraldehyde followed by osmium tetroxide, or (inset) with osmium tetroxide only. The continuities between the transverse tubular (or T) system and the sarcolemma are preserved by glutaraldehyde fixation (arrow). The T system appears in the form of small dense tubules after osmium tetroxide fixation and the continuities with the sarcolemma are disrupted, apparently by the action of the fixative. Z, Z line; SR, sarcoplasmic reticulum. (From Franzini-Armstrong and Porter 1964.)

Fig. 2.6.  Transverse sections of fish muscle fibres fixed with glutaraldehyde followed by osmium tetroxide, or (inset) with osmium tetroxide only. The circles mark areas of the A bands where the hexagonal arrays of thick and thin filaments are evident. The thick filaments are separated by the same distance in both images but they are obviously larger and more prominent after fixation with glutaraldehyde. Bridges between the filaments are indicated by arrows. The asterisks can be ignored. (From Franzini-Armstrong and Porter 1964.)

significant modifications in their molecular structure and is unable to prevent the much more drastic changes that occur during subsequent treatment with osmium tetroxide (Lenard and Singer 1968). However, it is doubtful whether these molecular changes are necessarily accompanied by morphological effects detectable in electron micrographs of thin sections. From a study of model systems, Millonig and Marinozzi (1968) concluded that the time for the fixation of proteins by glutaraldehyde will be greatly reduced in the presence of monovalent or divalent ions or sucrose. This contrasts with the result of adding sucrose to osmium tetroxide solutions which leads to an inhibition of proper fixation, and suggests that glutaraldehyde is particularly suitable for the fixation of material from sucrose gradients.

Glutaraldehyde reacts with glycogen and after double fixation the density

Fig. 2.7.   Myelinic figures (arrows) are associated with mitochondria in a lymphocyte fixed with glutaraldehyde followed by osmium tetroxide. (Unpublished micrograph from a study by Audrey M. Glauert and G. M. W. Cook.)

of glycogen is much higher than after fixation with osmium tetroxide alone, possibly as a result of interactions between osmium and the glutaraldehyde already bound to the glycogen (Millonig and Marinozzi 1968). Lipids are not preserved by glutaraldehyde and unless a second fixation with osmium tetroxide is used as much as 95% of the lipid in a specimen may be extracted during dehydration (Stein and Stein 1971). More seriously, glutaraldehyde and other aldehydes appear to cause phospholipids to pass into solution and myelinic figures (Fig. 2.7) then form when osmium tetroxide interacts with these phospholipids during the second fixation (Curgy 1968). These myelinic figures are seen more frequently in larger blocks of tissue. For- tunately, the formation of myelinic figures can be largely prevented by the addition of 1 to 3 mM calcium chloride to the glutaraldehyde fixative. Cal- cium also greatly reduces the loss of lipids during dehydration (Mitchell 1969) and improves the preservation of mitochondria (Busson-Mabillot 1971) (Fig. 2.8). As with osmium tetroxide fixatives, care must be taken not to

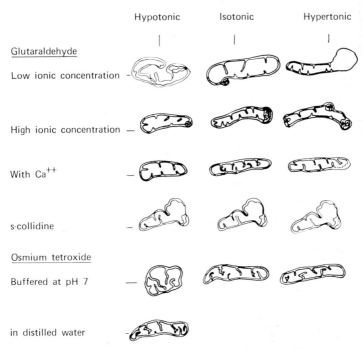

Fig. 2.8.   Variations in the fine structure of mitochondria under different conditions of fixation. The addition of low concentrations (0.015–0.05%) of calcium to the fixative reduces swelling and the formation of myelinic figures. The mitochondria appear very swollen after fixation with collidine-buffered glutaraldehyde. (From Busson-Mabillot 1971.)

add too much calcium as this will cause a precipitation of proteins (Millonig and Marinozzi 1968). If problems arise magnesium should be tested as an alternative (§ 2.2.4 and § 2.3.1).

Glutaraldehyde fixation does not destroy the osmotic properties of cells (e.g. Jard et al. 1966), although the cells are frequently changed so that subsequently they are stable only in solutions that are hypertonic or hypotonic relative to the original suspending medium. Glutaraldehyde itself contributes little to the osmotic effects of the fixative and volume changes are not detectable on passing the specimen from the fixative to the same solution without the glutaraldehyde (Bone and Denton 1971). Consequently, the osmolarity of the vehicle determines the effective osmolarity of the fixative. On this basis the majority of the glutaraldehyde fixatives used in electron microscopy are hypotonic, although they are usually described as being hypertonic since the glutaraldehyde itself is assumed to have an osmotic effect (Bone and Ryan 1972). This underlines yet again the impossibility of 'calculating' the 'correct' osmolarity for a fixative or depending on measurements of freezing-point depression.

Glutaraldehyde alone is not an adequate fixative for general use, since certain cell components, especially lipids, are not fixed and may be extracted during dehydration. Consequently, only the results of double fixation with glutaraldehyde followed by osmium tetroxide will be considered here (Fig. 2.9); the appearance of tissues fixed with glutaraldehyde alone was described by Sabatini et al. (1963) in their original paper.

2.4.2  GLUTARALDEHYDE SOLUTIONS

Glutaraldehyde is available commercially, usually as a 25% or 50% solution in water (see Appendix). Most of these solutions contain an impurity which shows spectrophotometric absorption at 235 nm, as compared with 280 nm for monomeric glutaraldehyde, and which has been postulated to be a polymer of glutaraldehyde. It is generally agreed that the monomeric glutaraldehyde is the main reactive species (Gillett and Gull 1972; Korn et al. 1972). In practice commercial solutions of glutaraldehyde, particularly Polysciences 'Biological Grade', which are relatively free of the impurity (Gillett and Gull 1972) have been found to be perfectly adequate for the majority of studies of fine structure. The solutions should be stored at 4 °C or at − 20 °C and barium carbonate should *not* be added. They should be discarded if the pH falls below 3.5 (Sabatini et al. 1964).

For certain special applications, such as cytochemical or immunological studies (see Lewis et al. 1974), it may be necessary to use glutaraldehyde

Fig. 2.9. An example of glutaraldehyde, followed by osmium tetroxide, fixation. An electron micrograph of a thin section showing good preservation of rough and smooth membranes, ribosomes and mitochondria in the cytoplasm of an osteoblast. (Unpublished micrograph from a study by Audrey M. Glauert, Honor B. Fell and J. T. Dingle.)

purified by charcoal filtration or distillation. Vacuum distilled glutaraldehyde is available commercially, usually as a 50% solution in water, and is stable for many months if stored at $-20\,°C$. Some commercial samples have a milky precipitate but this is easily removed by filtration (Fahimi and Drochmans 1968).

Vacuum distilled glutaraldehyde is very expensive and so it may be worthwhile purifying it in the laboratory shortly before use. In the simple method described by Smith and Farquhar (1966) the solution is distilled at atmospheric pressure and the distillate collected between 100 and 101 °C in 50 ml aliquots. Samples with a pH of less than 3.4 are discarded and the remainder pooled and bottled. Alternatively, the glutaraldehyde is distilled under vacuum at 15 mm Hg pressure, and the distillate is collected at 80–85 °C and immediately diluted to 25% with boiling water (Gillett and Gull 1972). A milky precipitate may form when the glutaraldehyde is diluted, but this can be prevented by warming the glutaraldehyde at 100 °C for 30 min before dilution (Fahimi and Drochmans 1968).

For purification with charcoal, a sample of commercial 25% glutaraldehyde is shaken with 10% (w/v) activated charcoal at 4 °C for 1 hr and then filtered. The process is repeated 2 or 3 times until a satisfactory UV absorption curve is obtained (Gillett and Gull 1972).

Gillett and Gull (1972) studied the conditions of storage of purified glutaraldehyde and found that temperature was the most important factor. No impurity formed after 8 months at $-20\,°C$ and only a small 235 nm peak was observed after 8 months at 4 °C, while large peaks developed after a relatively short time at room temperature and above. Storage in an inert atmosphere (nitrogen) or in the dark had little effect on the rate of deterioration.

The exact concentration of aldehyde groups in commercial samples of glutaraldehyde is not known, but an accurate knowledge is not usually required. If necessary, a measurement of optical density at 280 nm is a simple method of determining the concentration, as well as the osmolarity, of distilled glutaraldehyde (Fahimi and Drochmans 1965a) so long as it is remembered that the absorption is temperature dependent. Alternatively the method devised by Frigerio and Shaw (1969), which is based on the formation of a glutaraldehyde–bisulphite complex followed by iodometric titration of unreacted bisulphite, may be used. This method has a precision of 0.2% or better and is not affected by the presence of buffers so that it can be used directly with fixatives and does not require the distillation of the glutaraldehyde.

### 2.4.3 PREPARATION OF GLUTARALDEHYDE FIXATIVES

Since cells remain osmotically active during glutaraldehyde fixation the osmolarity of the fixative must be chosen carefully. This is done by adjusting the osmolarity of the fixative vehicle and not of the fixing agent (the glutaraldehyde) which has almost no osmotic effect (§ 2.4.1). The majority of fixatives used in electron microscopy, both for fixation by perfusion and by immersion, are hypotonic (if the effect of the glutaraldehyde is ignored), although they are often stated to be hypertonic since the glutaraldehyde is assumed to contribute to the osmolarity (Bone and Ryan 1972).

Again no general rule can be made in the choice of fixative and the beginner should first follow the procedures used successfully by others with similar specimens and then make modifications if signs of swelling or shrinkage are observed. The osmolarity of glutaraldehyde fixatives is adjusted by altering the osmolarity of the buffer or by the addition of sucrose, glucose or sodium chloride; sucrose does not appear to have any adverse effect on the fixation of proteins by glutaraldehyde (§ 2.4.1) and is generally used.

Although the osmolarity of glutaraldehyde fixatives is important, variations of pH within the range 6.8 to 7.5 rarely appear to have any effect. The pH should be kept below 7.5 to prevent polymerization of the glutaraldehyde and the loss of reactive groups (Sabatini et al. 1964), and for some tissues better fixation is obtained at pH 6.9 to 7.0 than at pH 7.2 to 7.4 (Busson-Mabillot 1971). Veronal-acetate is not a suitable buffer for aldehydes (§ 2.2.7) and collidine gives poor fixation with some tissues (§ 2.2.8). Results with phosphate and cacodylate buffers are very similar, so long as the osmolarity is the same, and a phosphate buffer should be used whenever possible (§2.2.1). It is strongly recommended that a small amount of calcium or magnesium (usually at a final concentration of 1 to 3 mM) be added to glutaraldehyde fixatives to minimise the formation of myelinic figures and other artefacts. Variations in the concentration of glutaraldehyde between 1.0 and 3.5% appear to have little effect (e.g. Fahimi and Drochmans 1965b).

Glutaraldehyde fixatives should be prepared and stored in clean, glass-stoppered bottles. Freshly prepared fixatives should be used whenever possible, although fixatives are stable for some weeks if kept at 4 °C in the dark in a refrigerator. The pH of cacodylate buffered glutaraldehyde may drop slightly during storage and should be tested regularly, while precipitates form in some phosphate buffered fixatives.

### Method of preparation of phosphate buffered glutaraldehyde fixative I

1. Prepare a 0.2 M phosphate buffer at the required pH (see § 2.2.5).
2. Prepare the fixative with:

| | | | | | |
|---|---|---|---|---|---|
| 0.2 M phosphate buffer (ml) | 50 | 50 | 50 | 50 | 50 |
| 25% glutaraldehyde in $H_2O$ (ml) | 4 | 6 | 8 | 10 | 12 |
| distilled water to make (ml) | 100 | 100 | 100 | 100 | 100 |
| final concentration of glutaraldehyde | 1.0% | 1.5% | 2.0% | 2.5% | 3.0% |

The pH may drop slightly on adding the glutaraldehyde. The final molarity of the buffer in the fixative is 0.1 M; fixatives of higher osmolarity may be prepared by starting with a more concentrated phosphate buffer.

3. If necessary, adjust the osmolarity of the fixative with sucrose, glucose or sodium chloride.
4. If required, add sufficient anhydrous calcium or magnesium chloride to give a final concentration of 1 to 3 mM, taking care to avoid the formation of a precipitate.

   (*Note:* tests for precipitation must be done at the temperature at which the fixative is to be used.)

### Method of preparation of phosphate buffered glutaraldehyde fixative II

(Karlsson and Schultz 1965; Maunsbach 1966a)

Prepare the fixative with

| | | | | |
|---|---|---|---|---|
| $NaH_2PO_4$. $H_2O$ (g) | 2.98 | 3.31 | 1.8 | 1.8 |
| $Na_2HPO_4$. $7H_2O$ (g) | 30.40 | 33.71 | 23.25 | 23.25 |
| NaCl (g) | – | – | 5.0 | 5.0 |
| 25% glutaraldehyde in water (ml) | 40 | 40 | 40 | 40 |
| Distilled water to make (ml) | 1000 | 1000 | 1000 | 2000 |
| Final concentration of glutaraldehyde | 2.5% | 2.5% | 2.5% | 1.25% |
| Osmolarity of buffer (mosmols) | 298 | 320 | 440 | 220 |
| pH of the fixative | 7.2–7.3 | 7.4 | 7.4 | 7.4 |

### Method of preparation of phosphate buffered glutaraldehyde fixative III

Prepare the fixative with

| | |
|---|---|
| 2.26% $NaH_2PO_4$. $H_2O$ in water | 64 ml |
| 25% glutaraldehyde in water | 8 ml |
| 2.52% NaOH in water, to adjust pH to required value | $x$ ml |
| distilled water to make | 100 ml |

The final concentration of glutaraldehyde is 2%.

### Method of preparation of cacodylate buffered glutaraldehyde fixatives

1. Prepare a 0.2 M cacodylate buffer at the required pH (see § 2.2.6).
2. Prepare the fixative with

| | | | | | |
|---|---|---|---|---|---|
| 0.2 M cacodylate buffer (ml) | 50 | 50 | 50 | 50 | 50 |
| 25% glutaraldehyde in water (ml) | 4 | 6 | 8 | 10 | 12 |
| distilled water to make (ml) | 100 | 100 | 100 | 100 | 100 |
| final concentration of glutaraldehyde | 1.0% | 1.5% | 2.0% | 2.5% | 3.0% |

The pH may drop slightly on adding the glutaraldehyde. The final molarity of the buffer is 0.1 M; fixatives of higher osmolarity may be prepared by starting with a more concentrated cacodylate buffer.

3. If necessary, adjust the osmolarity of the fixative with sucrose, glucose or sodium chloride.
4. If required, add sufficient calcium or magnesium chloride to give a final concentration of 1 to 3 mM calcium chloride in the fixative.

The pH of the buffered fixative may drop slightly during storage and should be checked before use.

### Method of preparation of collidine buffered glutaraldehyde fixatives

The fixative is prepared in exactly the same way as cacodylate buffered fixatives, with the substitution of 0.2 M collidine buffer for the 0.2 M cacodylate buffer.

## 2.5   Formaldehyde fixatives

Formaldehyde was tested as a fixative in the early days of electron microscopy but was found to give poorer results than osmium tetroxide and so
was little used, except as part of the 'formal-calcium-post-chromation'
procedure developed by Baker (1946) for the preservation of lipids (see
§ 3.9.1). Poor results were also obtained when formaldehyde was used as a
primary fixative before osmium tetroxide (Holt and Hicks 1961) due to the
presence of considerable amounts of methanol (11 to 16%) in commercial
solutions of formaldehyde (Pease 1964). Improved results were obtained with
methanol-free formaldehyde prepared from paraformaldehyde (Robertson
et al. 1963), which will be referred to as 'paraformaldehyde', although
formaldehyde is the actual fixing agent. Even better fixation of some
specimens results from the use of a mixture of paraformaldehyde and glutaraldehyde (Karnovsky 1965). Paraformaldehyde by itself is now only used
for certain special applications where its faster rate of penetration and different reactivity make it preferable to glutaraldehyde. As with other aldehydes
the fixation of lipids and membranes is improved by the addition of calcium
chloride, and fixation with paraformaldehyde must be followed by a second
fixation with osmium tetroxide. Specimens have a similar appearance to
those fixed with glutaraldehyde (Trump and Ericsson 1965).

### 2.5.1   CHEMICAL AND MORPHOLOGICAL EFFECTS OF FORMALDEHYDE

Formaldehyde reacts with proteins but there is less cross-linking than with
glutaraldehyde since formaldehyde is mono-functional, being a mono-
aldehyde. Also the reaction is slow and reversible (Hayat 1970) so that care
has to be taken during washing, and specimens cannot be stored for long
periods after formaldehyde fixation before further processing. Formaldehyde
reacts with lipids, although the chemistry of the reaction is not understood
(Millonig and Marinozzi 1968). It does not react with most polysaccharides,
but some glycogen is retained if the fixation is not prolonged.

The main advantage of formaldehyde is that it has a considerably higher
rate of penetration than either glutaraldehyde or osmium tetroxide so that
large blocks of tissue are well fixed. The permeability properties of membranes are partly preserved during formaldehyde fixation (Jard et al. 1966)
and the osmolarity of the fixative must therefore be adjusted carefully.
There is evidence that formaldehyde itself is not osmotically active (e.g.
Bernard and Wynn 1964).

### 2.5.2 FORMALDEHYDE SOLUTIONS

Commercially available solutions of formaldehyde are unsuitable for electron microscopy because of their content of methanol. Formaldehyde solutions are therefore prepared in the laboratory from powdered paraformaldehyde (trioxy-methylene) just before use.

***Method of preparation of methanol-free formaldehyde from paraformaldehyde***

**Note:** It is essential to work in a fume cupboard.

Prepare a 40% solution of paraformaldehyde by dissolving 40 g of para-formaldehyde powder in 100 ml of double distilled water by heating to 65 °C with continuous stirring. Add a few drops of 40% sodium hydroxide and the solution will become clear. The resulting solution of 40% para-formaldehyde is allowed to cool before mixing with the other components of the fixative.

### 2.5.3 PREPARATION OF PARAFORMALDEHYDE FIXATIVES

Paraformaldehyde fixatives have similar properties to glutaraldehyde fixatives (see § 2.4.3) and so the osmolarity should be adjusted carefully to a value suitable for the particular tissue under study. The choice of buffer is more important than with glutaraldehyde fixatives. Formaldehyde penetrates rapidly but reacts slowly and consequently there is a possibility that specific ions will have an effect before fixation is complete (e.g. Carson et al. 1972). Phosphate buffers should be used if possible, or cacodylate buffers, and a small amount of calcium chloride added. Collidine buffers are not suitable for general use since they cause lysis and extensive membrane destruction (Carson et al. 1972) (Fig. 2.3 and Table 2.1). The fixative should be prepared and stored in a clean, glass-stoppered bottle and is stable for some weeks if kept at 4 °C in the dark in a refrigerator. The pH of the fixative may drop slightly with time and should be tested before use.

***Method of preparation of paraformaldehyde fixatives I***

1. Prepare a 0.2 M phosphate (or cacodylate) buffer at the required pH (see § 2.2.5 and § 2.2.6).
2. Prepare the fixative with:

| | | | | |
|---|---|---|---|---|
| 0.2 M buffer (ml) | 50 | 50 | 50 | 50 |
| 40% paraformaldehyde in water (ml) | 2.5 | 5 | 7.5 | 10 |
| distilled water to make (ml) | 100 | 100 | 100 | 100 |
| final concentration of paraformaldehyde | 1.0% | 2.0% | 3.0% | 4.0% |

The pH may drop slightly on adding the paraformaldehyde. The final molarity of the buffer is 0.1 M; fixatives of higher molarity are prepared by starting with a more concentrated buffer.

3. If necessary, adjust the osmolarity of the fixative with sucrose, glucose or sodium chloride

4. If required, add sufficient calcium or magnesium chloride to give a final concentration of 1 to 3 mM, taking care to avoid the formation of a precipitate with a phosphate buffer.

### *Method of preparation of paraformaldehyde fixatives II*

Alternatively the fixative is prepared by dissolving the paraformaldehyde directly in the buffer.

1. Add 2.5 g of powdered paraformaldehyde to 50 ml of 0.2 M buffer, and heat to 65 °C with stirring. The resultant solution is cloudy.

2. Add a few drops of 1 N NaOH, with continued stirring, until the solution becomes clear.

3. Allow the solution to cool and then make up to 100 ml with distilled water.

The fixative contains 2.5% paraformaldehyde in 0.1 M buffer. The pH of the buffer will change as the paraformaldehyde dissolves. For example, Millonig and Bosco (1967) found that the pH of a phosphate buffer fell from 7.8 to 7.4 and then remained constant, while the pH of a cacodylate buffer rose from 7.2 to 7.4 and then fell to pH 7.0 after one week. Millonig and Bosco (1967) also found that not all commercial samples of para-formaldehyde dissolve in buffers and that some raise the pH to 10.0 or more.

### 2.5.4  PARAFORMALDEHYDE FIXATIVES FOR VERY LARGE TISSUE BLOCKS

In exceptional circumstances, such as in the fixation of surgical biopsies, very large blocks of tissue have to be fixed. Even fixatives containing a high concentration of paraformaldehyde do not penetrate sufficiently rapidly unless some agent is added to reduce the density of the tissue. A simple solution to this problem is to use a collidine buffered fixative since the extractive effects of collidine then become an advantage. Winborn and Seelig (1970) tested a collidine buffered 10% paraformaldehyde fixative and obtained adequate penetration of large blocks of tissue approximately 3.5 cm$^3$ in size.

## Method of preparation of collidine buffered paraformaldehyde fixatives

(Winborn and Seelig 1970)

1. Add 10 g of powdered paraformaldehyde to 70 ml of distilled water and heat the solution to 70 °C for 20 min with stirring. The resultant mixture is cloudy.
2. While the solution is still warm, add approximately 6 drops of 1 N NaOH with constant stirring. The solution will become clear.
3. Allow the solution to cool, add then add 2.4 ml of purified *s*-collidine.
4. Add 5 ml of 1 N HCl to adjust the pH to 7.3 to 7.4.
5. Make the solution up to 100 ml with distilled water.

The fixative contains 10% paraformaldehyde in 0.2 M collidine buffer.

### 2.5.5 PARAFORMALDEHYDE–GLUTARALDEHYDE FIXATIVES

Fixatives containing both paraformaldehyde and glutaraldehyde give better preservation of a wide variety of tissues than either aldehyde alone. In consequence, they are very widely used as primary fixatives at the present time (see Chapter 3). Formaldehyde penetrates tissues much more rapidly than glutaraldehyde and it is thought that the formaldehyde temporarily stabilises structures which are subsequently fixed more permanently by the glutaraldehyde (Karnovsky 1965). In consequence pieces of tissue are well fixed to a much greater depth and myelinic figures are less frequently seen than with glutaraldehyde alone. It is still advisable to add calcium chloride to the fixative whenever possible, and fixation must be followed by a second fixation with osmium tetroxide.

The fixative originally suggested by Karnovsky (1965) consists of 4% paraformaldehyde, 5% glutaraldehyde and 0.05% calcium chloride (approx. 5 mM) in approximately 0.08 M cacodylate buffer, pH 7.2, and is extremely hypertonic (2010 mosmols) if it is assumed that the aldehydes contribute to the osmolarity. Consequently lower concentrations of paraformaldehyde (0.5 to 2.0%) and glutaraldehyde (1.0 to 3.0%) are now used, the lowest concentrations being recommended for fixation by perfusion. However, the evidence that the aldehydes are not themselves osmotically active (§§ 2.4.1 and 2.5.1) raises doubts as to whether a reduction in osmolarity is the cause of the improved fixation. Some workers use distilled glutaraldehyde but there is no evidence that this is essential.

*Method of preparation of paraformaldehyde-glutaraldehyde fixatives*

1.  Prepare a 0.2 M phosphate or a 0.2 M cacodylate buffer (see § 2.2.5 and § 2.2.6) at the required pH. Cacodylate buffer is commonly used at pH 7.4.

2.  Prepare 20 ml of a 10% solution of paraformaldehyde by dissolving 2.0 g of paraformaldehyde powder in 20 ml of double distilled water and heating to 60 to 65 °C (in a fume cupboard). Add a few drops of 1.0 N sodium hydroxide until the solution becomes clear. Allow the solution to cool before use.

3.  Prepare the fixative with

    | | |
    |---|---|
    | 0.2 M buffer | 50 ml |
    | 10% paraformaldehyde in water | 20 ml |
    | 25% glutaraldehyde in water | 10 ml |
    | distilled water to make | 100 ml |

    This fixative contains 2% paraformaldehyde and 2.5% glutaraldehyde in 0.1 M buffer. Other concentrations are obtained by using different quantities of the constituents. The pH may change on adding the aldehydes.

4.  If necessary, adjust the osmolarity of the fixative with sucrose, glucose or sodium chloride.

5.  If required, add sufficient calcium or magnesium chloride to give a final concentration of 1 to 3 mM, taking care to avoid the formation of a precipitate with a phosphate buffer.

## 2.6   Acrolein and other aldehyde fixatives

### 2.6.1   ACROLEIN FIXATIVES

Acrolein (acrylic aldehyde) is the most rapidly penetrating of the aldehyde fixatives tested for electron microscopy (Sabatini et al. 1964) and is therefore useful for the fixation of large pieces of tissue and for plant cells and micro-organisms with dense cell walls. It is extremely toxic and must be handled with great care. It solubilises lipids, destroys most enzymatic activities, and causes loss of some microtubules (Schultz and Case 1968), and is only used when formaldehyde is unsuitable. As with other aldehyde fixatives, fixation must be followed by a second fixation with osmium tetroxide.

The acrolein available commercially (see Appendix) is a liquid and is suitable for immediate use. If necessary, the acrolein may be purified by distillation to remove the stabiliser (hydroquinone) and polymerized material

(Carstensen et al. 1971), but there is no evidence that purified acrolein gives better fixation. Acrolein fixatives are usually prepared with 10% acrolein in 0.025 or 0.05 M phosphate buffer. The fixatives are hypertonic and sucrose is not added.

### Method of preparation of acrolein fixatives

1. Prepare a 0.2 M phosphate or cacodylate buffer (see § 2.2.5 and § 2.2.6).
2. Prepare the 10% acrolein fixative with

| | | | |
|---|---|---|---|
| 0.2 M buffer (ml) | 50 | 25 | 12.5 |
| acrolein (ml) | 10 | 10 | 10 |
| distilled water to make (ml) | 100 | 100 | 100 |
| final concentration of buffer | 0.1 M | 0.05 M | 0.025 M |

3. If required, add sufficient calcium or magnesium chloride to give a final concentration of 1 to 3 mM taking care to avoid the formation of a precipitate with a phosphate buffer.

### 2.6.2 ACROLEIN–GLUTARALDEHYDE FIXATIVES

Acrolein is also used in combination with glutaraldehyde as an alternative to paraformaldehyde–glutaraldehyde fixatives (§ 2.5.5), and gives similarly excellent results.

### Method of preparation of acrolein-glutaraldehyde fixatives

1. Prepare a 0.2 M phosphate or cacodylate buffer (see § 2.2.5 and § 2.2.6) at the required pH.

2. Prepare the fixative with:

| | |
|---|---|
| 0.2 M buffer | 50 ml |
| acrolein | 1 ml |
| 25% glutaraldehyde in water | 10 ml |
| distilled water to make | 100 ml |

This fixative contains 1% acrolein and 2.5% glutaraldehyde in 0.1 M buffer. Other concentrations are obtained by using different quantities of the components. The pH may change on adding the aldehydes.

3. If necessary, adjust the osmolarity of the fixative with sucrose, glucose or sodium chloride.
4. If required, add sufficient calcium or magnesium chloride to give a final concentration of 1 to 3 mM taking care to avoid the formation of a precipitate with a phosphate buffer.

### 2.6.3  ACROLEIN–PARAFORMALDEHYDE–GLUTARALDEHYDE FIXATIVES

Hayat (1970) reports that fixation with a mixture of acrolein, paraformaldehyde and glutaraldehyde gives good fixation of a wide variety of animal and
plant tissues, and is especially effective for very dense tissues. The fixative
contains 10% acrolein (3 ml), 6% paraformaldehyde (5 ml), 10% glutaraldehyde (6 ml), 0.2 M buffer (5 ml) and distilled water (1 ml). It is prepared
by making a fresh solution of 6% paraformaldehyde and then adding the
other components.

### 2.6.4  OTHER ALDEHYDE FIXATIVES

A number of other aldehydes were tested as fixatives for electron microscopy by Sabatini et al. (1963) in their original study, including glyoxal,
hydroxyadipaldehyde, crotonaldehyde, pyruvic aldehyde, acetaldehyde and
methacrolein, and later the properties of malonaldehyde, malialdehyde and
succinaldehyde were examined also (Sabatini et al. 1964). None of them give
as good preservation of fine structure as glutaraldehyde, but some of them
(e.g. hydroxyadipaldehyde) are useful for enzyme cytochemical studies
(Lewis et al. 1974) since they are less reactive than glutaraldehyde and have
little effect on the activity of many enzymes.

Acetaldehyde must be handled with particular care since it is very volatile
and has a pungent vapour. It boils at 20 °C and must be stored and used at
0 °C or below.

## 2.7  Permanganate fixatives

### 2.7.1  POTASSIUM PERMANGANATE FIXATIVES

Potassium permanganate was suggested as an alternative fixative to osmium
tetroxide by Luft (1956) and was popular for the fixation of botanical
specimens and membrane systems in the period before the introduction of
aldehyde fixation by Sabatini et al. (1963). Permanganate fixatives were
favoured for membranes (e.g. Robertson 1959), since they stand out with
great clarity against the surrounding cytoplasm. It was later appreciated
that this is a consequence of the extraction of many cell components,
including soluble cytoplasmic proteins, ribosomes, neurofilaments, neuro-
tubules and myofilaments, during fixation and the subsequent dehydration.
These destructive effects are better understood now that more is known
about the interactions between potassium permanganate and cell components
(Hayat 1970; Riemersma 1970). It has been found that potassium permanganate does not stabilise protein gels in model experiments (Bradbury and

Meek 1960), obliterates most of the helical character of isolated proteins in solution (Lenard and Singer 1968) and extracts a large part of the proteins from cells during fixation (Millonig and Marinozzi 1968). Furthermore fixation with potassium permanganate does not prevent the extraction of large amounts of lipid from rat liver (Cope and Williams 1969) or nerve myelin (Moretz et al. 1969) during dehydration, more lipid being extracted than after osmium tetroxide fixation, and comparative studies have shown that some components of membranes visible after osmium tetroxide fixation are no longer detectable after potassium permanganate fixation (Sjöstrand and Elfvin 1962).

RNA is completely extracted from tissues while DNA is preserved (Bradbury and Meek 1960) although potassium permanganate is not a good fixative for chromosomes (Porter and Machado 1960). Glycogen is fixed and stained, but the chemical basis of the reaction has not yet been defined (Millonig and Marinozzi 1968).

The inability of potassium permanganate to preserve many cellular components and the swelling induced in many organelles (Fig. 2.10) makes it unsuitable as a general fixative for electron microscopy. It is still favoured, however, for some botanical studies where its ability to penetrate thick cell walls is an advantage (Van Steveninck 1972). It is of value in the examination of some membrane systems since comparative studies have shown that the continuity of certain membranes is better preserved by potassium permanganate than by other fixatives (Doggenweiler and Heuser 1967; Martin and Rosenberg 1968) although in general the reverse is true. Such observations emphasise the importance of comparing results with different procedures so that systematic artefacts can be detected (see Chapter 1).

Luft (1956) originally suggested the use of a veronal–acetate buffer (§ 2.2.7) for potassium permanganate fixatives, but for many tissues the results are no worse if an unbuffered solution is used (e.g. Mollenhauer 1959). The time of fixation with potassium permanganate often appears to be critical.

### *Method of preparation of potassium permanganate fixatives*
(Luft 1956)

1. Prepare a veronal–acetate stock solution with

|  |  |  |
|---|---|---|
| | sodium veronal (barbitone sodium) | 2.89 g |
| | sodium acetate (anhydrous) | 1.15 g |
| *or* | sodium acetate (hydrated) | 1.90 g |
| | distilled water to make | 100 ml |

This solution is stable for some months at 4 °C.

Fig. 2.10.   Section of a rabbit kidney (RK 13) cell fixed with potassium permanganate. Membranes are clearly visible but the mitochondria appear swollen and the ribosomes have not been preserved. (From Dimmock 1970.)

2. Prepare a 1.2% solution of potassium permanganate with

        potassium permanganate ($KMnO_4$)     1.2 g

        distilled water to make         100  ml

This solution is slowly reduced on contact with air and should be stored in a well-filled, glass-stoppered bottle in a refrigerator.

3. Prepare the potassium permanganate fixative with

        1.2% potassium permanganate     12.5 ml

        veronal-acetate stock solution     5.0 ml

        distilled water         2.5 ml

        0.1 NHCl         5.0 ml

The potassium permanganate solution, veronal–acetate stock solution and distilled water are mixed together in a brown, glass-stoppered bottle, and then the HCl is added gradually until the required pH (usually 7.2 to 7.6) is reached. The final concentration of potassium permanganate is 0.6%; higher concentrations, up to 5%, have been used by some workers and are obtained by starting with a more concentrated solution of potassium permanganate. The fixative is not stable and must be prepared freshly just before use. Millonig and Marinozzi (1968) suggest that 0.5% sodium chloride be added to the fixative to prevent swelling of isolated mammalian cells.

### 2.7.2 SODIUM PERMANGANATE FIXATIVES

Although potassium permanganate fixation reveals membranes with great clarity, some membranes are disrupted or even destroyed. The observation that plasma membranes of cells in free-living flat worms show discontinuities after potassium permanganate fixation led Wetzel (1961) to test sodium permanganate as an alternative fixative. He stressed the need to maintain a delicate balance between sodium, potassium and calcium ions for the preservation of cell structure and function, and pointed out that disruption of membranes can occur even if the osmolarity of the fixative is correct (see § 2.2.3). The concentration of potassium ions in a 1% potassium permanganate solution greatly exceeds the values generally reported in body fluids, while 1% sodium permanganate does not increase the sodium concentration above that usually found. In a comparison of potassium permanganate and sodium permanganate fixation Wetzel (1961) found that the results obtained were quite similar, although with sodium permanganate there was a slightly, but significantly, better preservation of the integrity and continuity of plasma membranes. Sodium permanganate was equally disruptive to pigment granules and ribosomes.

### Method of preparation of sodium permanganate fixative I
(Wetzel 1961)

The 1% sodium permanganate fixative is prepared in exactly the same way as potassium permanganate fixatives (§ 2.7.1). Wetzel (1961) fixed specimens for 2 hr at 1–2 °C in a fixative buffered at pH 6.0.

### Method of preparation of sodium permanganate fixative II
### (for mammalian tissues)
(Rosenbluth 1963)

The ionic composition of Wetzel's fixative was further modified by Rosenbluth (1963) by the addition of salts to mimic the ionic composition of mammalian and amphibian body fluids.

1.  Prepare a solution of 0.83% $NaMnO_4$. $3H_2O$ in veronal–acetate buffer, pH 7.5, by the method used for potassium permanganate (§ 2.7.1).

2.  Prepare the sodium permanganate fixative with

| | |
|---|---|
| NaCl | 2.8 g |
| KCl | 0.4 g |
| $CaCl_2$ | 0.2 g |
| $MgCl_2$. $6H_2O$ | 0.2 g |
| $NaH_2PO_4$ | 0.16 g |
| 0.83% buffered sodium permanganate to make | 100   ml |

The fixative is not stable and is prepared freshly just before use.

### Method of preparation of sodium permanganate fixative III
### (for amphibian tissues)
(Rosenbluth 1963)

Sodium permanganate fixative II is diluted to 75% of its initial concentration with distilled water. The final concentration of sodium permanganate is about 0.62% and the fixative has the same concentration of sodium, potassium, calcium, magnesium and phosphate ions as an amphibian salt solution.

Rosenbluth (1963) fixed toad spinal ganglia for 1–2 hr at 5 °C and obtained better preservation of slender invaginations of the plasma membrane than with a similarly buffered osmium tetroxide fixative, but a less satisfactory fixation of thin sheets of cytoplasm. Reorganisation of membranes was observed and was thought to be due to physical instability created in the membrane following changes in surface proteins produced by the fixative.

### 2.7.3 LANTHANUM PERMANGANATE FIXATIVES

Doggenweiler and Frenk (1965) investigated the possibility of using lanthanum permanganate as a fixative for cell membranes on the assumption that $La^{3+}$ would provide an even greater stabilisation of membranes than $Ca^{2+}$. They found that lanthanum permanganate stains intercellular substances in vertebrate and invertebrate nervous tissue, and a surface layer of plasma membranes (Fig. 2.11). The mechanism of this staining by lanthanum is not yet understood (Lewis et al. 1974), but there is a warning in the study by Dimmock (1970) who showed that the staining is much denser in the presence of anions, such as phosphate, that can form insoluble salts with lanthanum. She recommends that specimens are washed in phosphate-free buffer solution before fixation with lanthanum permanganate.

Lanthanum does not penetrate into cells, and intracellular structures appear the same as after potassium permanganate fixation (Lesseps 1967; Dimmock 1970) (Fig. 2.11).

Doggenweiler and Frenk (1965) prepared lanthanum permanganate from lanthanum sulphate and barium permanganate. A simpler method is described by Lesseps (1967).

### *Method of preparation of lanthanum permanganate fixatives*
(Lesseps 1967)

1. Prepare a veronal-acetate stock solution with

   | | | |
   |---|---|---|
   | sodium veronal | 2.94 g | |
   | sodium acetate (hydrated) | 1.94 g | |
   | distilled water to make | 100 | ml |

   This solution is stable and keeps for some months at 4 °C.

2. Prepare a Ringer's solution (Zetterqvist 1956) with

   | | | |
   |---|---|---|
   | sodium chloride | 8.05 g | |
   | potassium chloride | 0.42 g | |
   | calcium chloride | 0.18 g | |
   | distilled water to make | 100 | ml |

Fig. 2.11.    Section of an RK 13 cell fixed with lanthanum permanganate. The appearance of the cell is similar to that obtained with potassium permanganate fixatives (cf. Fig. 2.10), except for the dense staining of a layer of material at the cell surface.
(From Dimmock 1970.)

3. Prepare the lanthanum permanganate fixative with

| | |
|---|---|
| $La(NO_3)_3 \cdot 6H_2O$ | 1 g |
| $KMnO_4$ | 1 g |
| veronal–acetate stock solution | 20 ml |
| Ringer's solution | 6 ml |
| 0.1 N HCl | $x$ ml to adjust pH to 7.8 |
| distilled water to make | 100 ml |

Fragments of tissue or isolated cells are fixed for 1 hr at 0 °C.

### 2.7.4 OTHER PERMANGANATE FIXATIVES

Comparative studies on the effects of fixation by a range of permanganates have shown that other monovalent salts (lithium and sodium) give results of comparable quality to potassium permanganate (Afzelius 1962, 1963; Hökfelt and Jonsson 1968). Membranes appear very distinct after fixation with calcium, barium or zinc permanganate, and this is thought to be due to the stabilising action of divalent cations on lipids. Cytoplasmic preservation is not good with zinc permanganate fixatives (Afzelius 1962) and some tissues are destroyed (Hökfelt and Jonsson 1968). Doggenweiler and Frenk (1965) reported that barium permanganate produced the same appearance in retina and sciatic nerve as potassium permanganate but with increased contrast.

## 2.8 Mixed fixatives

Improved preservation of a wider range of structures can often be obtained by mixing two fixing agents together and these mixed fixatives are considered in the present section. Equally good results can often be achieved by using a number of fixatives in sequence as described in the following chapter.

### 2.8.1 OSMIUM TETROXIDE–POTASSIUM DICHROMATE FIXATIVES

Potassium dichromate neutralised with potassium hydroxide is a strong buffer system in the pH range 5.6 to 7.6 and consequently Dalton (1955) proposed that a mixture of potassium dichromate and osmium tetroxide (chrome–osmium) be used as an alternative to Palade's (1952) veronal–acetate buffered osmium tetroxide fixative (§ 2.3.5). Improved results were obtained with a variety of tissues since there was less leaching out of cytoplasmic materials. This was probably due not only to the presence of a more effective buffer system, but also to the fact that the addition of potassium dichromate accelerates the penetration of osmium tetroxide (Riemersma 1970).

Since the introduction of aldehyde fixatives, chrome–osmium is mainly used as a second fixative as an alternative to osmium tetroxide alone.

### *Method of preparation of osmium tetroxide-potassium dichromate fixatives*

(Dalton 1955)

1. Prepare a buffered solution of 4% potassium dichromate in water with

   | | |
   |---|---|
   | 5% $K_2Cr_2O_7$ in water | 80 ml |
   | 2.5 N KOH | $x$ ml to adjust pH to 7.2 |
   | distilled water to make | 100 ml |

2. Prepare the fixative with

   | | |
   |---|---|
   | 4% potassium dichromate, pH 7.2 | 5.0 ml |
   | 2% osmium tetroxide (§ 2.3.2) | 10.0 ml |
   | 3.4% sodium chloride in water | 5.0 ml |

The fixative contains 1% potassium dichromate, 1% osmium tetroxide and 0.85% sodium chloride, and has a pH of 7.2.

The pH does not change on adding the osmium tetroxide. The fixative is stable for some months at 4 °C.

### 2.8.2   GLUTARALDEHYDE–OSMIUM TETROXIDE FIXATIVES

Neither glutaraldehyde nor osmium tetroxide is suitable as a general fixative when used alone, and certain artefacts are observed in some specimens when fixation with glutaraldehyde is followed by fixation with osmium tetroxide. These artefacts arise due to the deleterious effects of glutaraldehyde, such as lipid extraction and cell shrinkage, which occur before the osmium tetroxide is applied. Trump and Bulger (1966) showed that most of these effects can be minimised or avoided by simultaneous use of glutaraldehyde and osmium tetroxide in a mixed fixative. The fixative is not stable, since the two fixing agents react with each other. The mechanism of fixation is not known since each of the reagents may be in a different state than when used alone. For example, Hopwood (1970) has shown that the rate of cross-linking of proteins by glutaraldehyde is decreased in the presence of osmium tetroxide, due to competition for amino acid residues.

The fixative suggested by Trump and Bulger (1966) consisted of 6.25% glutaraldehyde and 1.0% osmium tetroxide in collidine buffer. This fixative was later modified by Hirsch and Fedorko (1968) who used a lower concentration of glutaraldehyde and a cacodylate buffer, and followed the

Fig. 2.12. Section of parts of two neighbouring C3 HS/1 cells in a monolayer culture fixed with a mixture of glutaraldehyde and osmium tetroxide. (Unpublished micrograph from a study by Audrey M. Glauert and Mary R. Daniel.)

glutaraldehyde-osmium tetroxide fixation with a second fixation with uranyl acetate (see § 2.9). The osmolarity of the fixative vehicle does not appear to be of importance (Bone and Ryan 1972).

### Method of preparation of glutaraldehyde-osmium tetroxide fixatives
  (Hirsch and Fedorko 1968)

1. Prepare a stock solution of 2.5% glutaraldehyde in 0.1 M cacodylate buffer, pH 7.4 (§ 2.4.3).
2. Prepare a stock solution of 10% osmium tetroxide in 0.1 M cacodylate buffer, pH 7.4 (§ 2.3.4).
3. Bring the two solutions to 0 °C in an ice bath, and mix one part of glutaraldehyde solution with 2 parts of osmium tetroxide solution.

The fixative consists of about 0.8% glutaraldehyde and 0.7% osmium tetroxide in 0.1 M cacodylate buffer, pH 7.4. The relative concentrations of glutaraldehyde and osmium tetroxide are not critical. Fixation must be carried out at 0 to 4 °C to reduce the rate of reaction between the glutaraldehyde and the osmium tetroxide, and specimens are fixed for a maximum of one hour. At room temperature the fixative turns red-brown after 1 hr and black after 6 hr, and the pH falls (Trump and Bulger 1966).

Cells and tissues fixed with the mixed fixatives have many cytological features intermediate in appearance between those seen with either fixative alone (Fig. 2.12). The mixture is recommended for fragile structures, such as those in pathological cells (Trump and Bulger 1966) and for isolated cells and monolayers of cells (Hirsch and Fedorko 1968).

Presumably mixtures of osmium tetroxide with other aldehydes would give similar results. A mixture of 2% osmium tetroxide and Karnovsky's paraformaldehyde–glutaraldehyde fixative (§ 2.5.5) was tested by Pollard and Ito (1970) and gave comparable preservation of amoebae to aldehyde fixatives containing trinitro compounds (§ 2.8.4).

### 2.8.3  FIXATIVES CONTAINING POTASSIUM FERRICYANIDE

The addition of potassium ferricyanide to fixatives for the preservation of lipids was first suggested by Elbers et al. (1965) in a study of the fixation of phospholipids by tricomplex 'flocculation'. Subsequently, potassium ferricyanide was added to various fixatives for the preservation of labile lipids, and densely stained material was observed in regions where lipids were thought to be present. Recently, however, Gil (1972) has shown that stained material is still visible after the extraction of phospholipids and concluded that no statements about the chemical nature of the reaction product can be made. In spite of these doubts, the fixatives appear to have some value in morphological studies and so are included here. Various fixation schedules have been used.

Dermer (1969) fixed lung tissue by immersion firstly in 2% glutaraldehyde in cacodylate buffer, pH 7.2, then in a mixture of 0.05 M potassium ferricyanide $(K_3Fe(CN)_6)$ and 0.05 M lead nitrate, and finally in an osmium tetroxide fixative. Finlay-Jones and Papadimitrou (1972) showed that better results were obtained if the initial fixation was done by perfusion.

Bluemink (1972) fixed *Xenopus* eggs by a method developed by De Bruijn (1969). The eggs were immersed firstly in 1% distilled acrolein plus 2.5% glutaraldehyde in 0.067 M cacodylate buffer, pH 7.4 containing 1 mM calcium chloride (for 24 hr at room temperature), then stored overnight in the dark in the same buffer containing calcium chloride, and finally fixed in 1% osmium tetroxide in the same buffer, containing 0.05 M potassium ferricyanide and 0.05 M calcium chloride (for $3\frac{1}{2}$ hr in the dark and cold). The method was successful in preserving labile lipids (Fig. 2.13).

Karnovsky (1971) reported that an osmium tetroxide–potassium ferricyanide fixative greatly increased the contrast of the outer surfaces of membranes and of glycogen.

### 2.8.4 ALDEHYDE FIXATIVES CONTAINING TRINITRO COMPOUNDS

The effect of adding trinitro compounds to an aldehyde fixative was examined by Stefanini et al. (1967) and Zamboni and De Martino (1967) who used a mixture of paraformaldehyde and trinitrophenol (picric acid) (PAF) for the fixation of spermatozoa and large pieces of tissue. The fixative penetrated rapidly and gave good preservation of whole small embryos and whole organs, such as the kidney. It was very stable, withstanding exposure to light at room temperature for 12 months. Subsequently Ito and Karnovsky (1968) explored the use of a range of trinitro compounds which were added at concentrations of 0.02 to 1.0% to the paraformaldehyde–glutaraldehyde fixative of Karnovsky (1965)(§ 2.5.5) diluted to contain 2% paraformaldehyde and 2.5% glutaraldehyde in 0.1 or 0.2 M phosphate or cacodylate buffer at pH 7.2. Five different trinitro compounds were tested and it was found that 2,4,6-trinitrocresol (FGC), 2,4,6-trinitroresorcinol (FGR) and 2,4,6-trinitrophenol (FGP) give markedly better fixation than 1,3,5-trinitrobenzene (FGB) and 2,4,6-trinitrotoluene (FGT), and that all these fixatives are superior to the paraformaldehyde–picric acid (PAF) fixative used by Stefanini et al. (1967).

**Note:** All trinitro compounds are potentially *explosive*.

Fig. 2.13a

Fig. 2.13b

Fig. 2.13. Sections of surface protrusions in the region of furrow formation in the first cleavage zygote of *Xenopus laevis*. (a) After fixation with glutaraldehyde and acrolein, followed by osmium tetroxide (Bluemink 1970) the membrane at the tip of a surface protrusion shades off into a bleb which is barely visible. (b) After fixation with glutaraldehyde and acrolein, followed by osmium tetroxide and potassium ferricyanide (Bluemink 1972) the bleb at the tip of a surface protrusion consists of material arranged in myelin-like configuration. (Unpublished micrographs from a study by John G. Bluemink.)

After fixation specimens must be washed very thoroughly in buffer (preferably overnight) to remove all the fixative before secondary fixation with osmium tetroxide; small black particles are observed in insufficiently washed tissues.

Ito and Karnovsky (1968) obtained more consistent results with FGC, FGR and FGP, with a variety of tissues fixed by immersion or perfusion, than with paraformaldehyde–glutaraldehyde alone. In particular, membrane systems were well fixed, myelin figures were rare with FGP and FGC, and excellent preservation of the smooth endoplasmic reticulum in steroid-secreting cells was obtained.

Three of these mixed fixatives (FGR, FGC and FGP) were also tested by Pollard and Ito (1970) for the fixation of amoebae and gave satisfactory results, FGC being preferred since it gave the fewest myelin figures. Problems were encountered, however, since the organisms shrank to as little as half their original volume during fixation. If the amoebae were passed rapidly through the fixative to avoid shrinkage, they then shrank during dehydration. In addition, Pollard and Ito (1970) also noted that the amoebae underwent violent abnormal contractions during the first few minutes in the fixative at 22 °C. These contractions did not occur in cold fixative and so it is essential to use the fixative at 0–4 °C.

### 2.8.5  FIXATIVES CONTAINING DIGITONIN

The addition of 0.2% digitonin to an aldehyde fixative helps to retain free cholesterol which is extracted following convential fixation procedures (Frühling et al. 1969; Scallen and Dietert 1969). This enhanced retention of cholesterol is accompanied by an improvement in the morphological preservation of some types of cell, except for the appearance of many small cylindrical 'spicules' in association with membranes. This artefact can be avoided by using a mixed fixative consisting of 2.5% glutaraldehyde, 1% osmium tetroxide and 0.2% digitonin in a cacodylate buffer at pH 7.4 (Napolitano et al. 1969).

### 2.8.6  OTHER ADDITIVES FOR FIXATIVES

There are many reports in the literatures of experiments on the effects of adding various compounds to aldehyde fixatives; most of these fixatives are used for special purposes and none of them appear to be suitable for general application. These compounds include potassium permanganate (Kaye and Moses 1960), a nitrogen mustard derivative (Williams and Luft 1968) and potassium dichromate (Robison and Lipton 1969).

Dimethyl sulphoxide has been added to aldehyde fixatives to accelerate penetration (Sandborn et al. 1969; Schwab et al. 1970), but may have a disruptive effect on membranes if the concentration is too high (Winborn and Seelig 1970).

A number of compounds which react with polysaccharides have been added to fixatives, particularly for the fixation and staining of cell surfaces. These include alcian blue (Behnke and Zelander 1970), lanthanum (e.g. Revel and Karnovsky 1967) and ruthenium red (Luft 1971).

Shea and Karnovsky (1969) and Shea (1971) added alcian blue or cetyl-pyridinium chloride to the primary glutaraldehyde fixative, and lanthanum nitrate to the secondary osmium tetroxide fixative and reported an enhance-ment of the staining obtained by the addition of lanthanum alone to the fixatives. A mixed fixative containing osmium tetroxide and zinc iodide appears to stain certain lipids (e.g. Blümcke et al. 1973), while glutaraldehyde treated with hydrogen peroxide gives improved fixation of a variety of tissues (Peracchia and Mittler 1972) (Fig. 2.14).

The staining properties of fixatives are discussed in detail by Lewis et al. (1974) in another book in this series.

## 2.9  Uranyl acetate fixatives

In a study of the fixation of bacteria Ryter and Kellenberger (1958) showed that a 'wash' in a solution of uranyl acetate in a veronal–acetate buffer, following osmium tetroxide fixation and preceding dehydration, helped to stabilise the nuclear material. Subsequently, treatment with uranyl acetate became a routine in the fixation of bacteria (§ 3.9.3).

More recently it has been realised that uranyl acetate is also useful as a fixative for animal tissues, being used as a third fixative following fixation with an aldehyde and with osmium tetroxide (e.g. Farquhar and Palade 1965). It is particularly effective in the fixation of membranes (e.g. Brightman and Reese 1969) (Fig. 2.15), as a consequence of the stabilisation of phospho-lipids (Silva et al. 1968; Hayat 1970; Silva et al. 1971), although there is an accompanying extraction of other components (Farquhar and Palade 1965). For example, glycogen may be extracted or rendered unstainable.

The fixative consists of 0.25 to 2.0% uranyl acetate in water or in a veronal–acetate (Ryter and Kellenberger 1958) or sodium maleate (Karnovsky 1967) buffer. The final pH is usually in the region of pH 5.0. Phosphate and cacodylate buffers cannot be used since precipitation occurs, and tissues fixed in phosphate or cacodylate buffered osmium tetroxide must be

Fig. 2.14.   Transverse section of fish muscle fixed with glutaraldehyde treated with hydrogen peroxide, followed by osmium tetroxide and then uranyl acetate. Compare with Fig. 2.6. (Unpublished micrograph from a study by Clara Franzini-Armstrong.)

Fig. 2.15. A comparison of the appearance of gap junctions after various preparative procedures. (a) Initial fixation in osmium tetroxide followed by immersion in potassium permanganate after dehydration. (b) Potassium permanganate fixation only. (c) Initial fixation in osmium tetroxide followed by immersion in uranyl acetate before dehydration. (d) Fixation in osmium tetroxide only. Treatment with uranyl acetate before dehydration gives the best preservation of membranes. (From Brightman and Reese 1969.)

washed thoroughly before fixation with uranyl acetate (Farquhar and Palade 1965; Hayat 1970). The fixative should be made up freshly about once a week, and should be stored at room temperature in the dark. The staining properties of the fixative decrease with time and a precipitate gradually forms. The staining effects of uranyl acetate are described by Lewis et al. (1974).

REFERENCES

Afzelius, B. A. (1962), Chemical fixatives for electron microscopy, in: The interpretation of ultrastructure, R. J. C. Harris, ed. (Academic Press, New York and London), p. 1.
Afzelius, B. A. (1963), Experiments with simple fixatives, J. Ultrastruct. Res. *9*, 393.
Amsterdam, A. and M. Schramm (1966), Rapid release of the zymogen granule protein by osmium tetroxide and its retention during fixation by glutaraldehyde, J. Cell Biol. *29*, 199.
Bahr, G. F., G. Bloom and U. Friberg (1957), Volume changes of tissues in physiological fluids during fixation in osmium tetroxide or formaldehyde and during subsequent treatment, Expl Cell. Res. *12*, 342.
Baker, J. R. (1946), The histochemical recognition of lipine, Q. Jl Microsc. Sci. *87*, 441.
Baker, J. R. (1965), The fine structure produced in cells by fixatives, Jl R. microsc. Soc. *84*, 115.
Behnke, O. and T. Zelander (1970), Preservation of intercellular substances by the cationic dye Alcian blue in preparative procedures for electron microscopy, J. Ultrastruct. Res. *31*, 424.
Bennett, H. S. and J. H. Luft (1959), s-collidine as a basis for buffering fixatives, J. biophys. biochem. Cytol. *6*, 113.
Bernard, G. R. and G. G. Wynn (1964), Weight responses of tissue slices and albumin–gelatin gels during formaldehyde fixation with observations upon the effect of pH, Anat. Rec. *150*, 463.
Bluemink, J. G. (1970), The first cleavage of the amphibian egg, J. Ultrastruct. Res. *32*, 142.
Bluemink, J. G. (1972), Cortical wound healing in the amphibian egg: an electron microscopical study, J. Ultrastruct. Res. *41*, 95.
Blümcke, S., W. D Kessler, H. R. Niedorf, N. H. Becker and F. J. Veith (1973), Ultra-

structure of lamellar bodies of type II pneumocytes after osmium–zinc impregnation, J. Ultrastruct. Res. *42*, 417.

Bone, Q. and E. J. Denton (1971), The osmotic effects of electron microscope fixatives, J. Cell Biol. *49*, 571.

Bone, Q. and K. P. Ryan (1972), Osmolarity of osmium tetroxide and glutaraldehyde fixatives, Hist. J. *4*, 331.

Bradbury, S. and G. A. Meek (1960), A study of potassium permanganate 'fixation' for electron microscopy, Q. Jl Microsc. Sci. *101*, 241.

Brightman, M. W. and T. S. Reese (1969), Junctions between intimately apposed cell membranes in the vertebrate brain, J. Cell Biol. *40*, 648.

Busson-Mabillot, S. (1971), Influence de la fixation chimique sur les ultrastructures. 1. Étude sur les organites du follicule ovarien d'un poisson téléostéen, J. Microscopie *12*, 317.

Carson, F., J. A. Lynn and J. H. Martin (1972), Ultrastructural effect of various buffers, osmolality, and temperature on paraformaldehyde fixation of the formed elements of blood and bone marrow, Tex. Rep. Biol. Med. *30*, 125.

Carstensen, E. L., W. G. Aldridge, S. Z. Child, C. P. Sullivan and H. H. Brown (1971), Stability of cells fixed with glutaraldehyde and acrolein, J. Cell Biol. *50*, 529.

Caulfield, J. B. (1957), Effects of varying the vehicle for $OsO_4$ in tissue fixation, J. biophys. biochem. Cytol. *3*, 827.

Claude, A. (1962), Fixation of nuclear structures by unbuffered solutions of osmium tetroxide in slightly acid distilled water, Proc. 5th Int. Congr. Electron Microscopy, Philadelphia *2*, L-14.

Cope, G. H. and M. A. Williams (1969), Quantitative studies on the preservation of choline and ethanolamine phosphatides during tissue preparation for electron microscopy. II. Other preparative methods, J. Microscopy *90*, 47.

Curgy, J.-J. (1968), Influence du mode de fixation sur la possibilité d'observer des structures myéliniques dans les hepatocytes d'embryons de poulet, J. Microscopie *7*, 63.

Dalton, A. J. (1955), A chrome–osmium fixative for electron microscopy, Anat. Rec. *121*, 281.

Dawson, R. M. C., D. C. Elliott, W. H. Elliott and K. M. Jones (1969), Data for biochemical research, 2nd edition (Clarendon Press, Oxford).

De Bruijn, W. C. (1969), Thesis, Rotterdam.

Dermer, G. B. (1969), The fixation of pulmonary surfactant for electron microscopy. 1. The alveolar surface lining layer, J. Ultrastruct. Res. *27*, 88.

Dimmock, E. (1970), The surface structure of cultured rabbit kidney cells as revealed by electron microscopy, J. Cell Sci. *7*, 719.

Doggenweiler, C. F. and S. Frenk (1965), Staining properties of lanthanum on cell membranes, Proc. Natn. Acad. Sci. U.S.A. *53*, 425.

Doggenweiler, C. F. and J. E. Heuser (1967), Ultrastructure of the prawn nerve sheath. Role of fixative and osmotic pressure in vesiculation of thin cytoplasmic laminae, J. Cell Biol. *34*, 407.

Elbers, P. F., P. H. J. T. Ververgaert and R. Demel (1965), Tricomplex fixation of phospholipids, J. Cell Biol. *24*, 23.

Fahimi, H. D. and P. Drochmans (1965a), Essais de standardisation de la fixation au glutaraldéhyde. I. Purification et détermination de la concentration du glutaraldéhyde, J. Microscopie *4*, 725.

Fahimi, H. D. and P. Drochmans (1965b), Essais de standardisation de la fixation au glutaraldéhyde. II. Influence des concentrations en aldéhyde et de l'osmolalité, J. Microscopie *4*, 737.

Fahimi, H. D. and P. Drochmans (1968), Purification of glutaraldehyde and its significance for preservation of acid phosphates activity, J. Histochem. Cytochem. *16*, 199.

Farquhar, M. G. and G. E. Palade (1965), Cell junctions in amphibian skin, J. Cell Biol. *26*, 263.

Finlay-Jones, J. M. and J. M. Papadimitrou (1972), Demonstration of pulmonary surfactant by tracheal injection of tricomplex salt mixture: electron microscopy, Stain technol. *47*, 59.

Franzini-Armstrong, C. and K. R. Porter (1964), Sarcolemmal invaginations constituting the T system in fish muscle fibers, J. Cell Biol. *22*, 675.

Frigerio, N. A. and M. J. Shaw (1969), A simple method for determination of glutaraldehyde, J. Histochem. Cytochem. *17*, 176.

Frühling, J., W. Penasse, G. Sand and A. Claude (1969), Préservation du cholostérol dans la corticosurrénale du rat au cours de la préparation des tissus pour la microscopie électronique, J. Microscopie *8*, 957.

Gil, J. (1972), Effect of tricomplex fixation on lung tissue, J. Ultrastruct. Res. *40*, 122.

Gil, J. and E. R. Weibel (1968), The role of buffers in lung fixation with glutaraldehyde and osmium tetroxide, J. Ultrastruct. Res. *25*, 331.

Gillett, R. and K. Gull (1972), Glutaraldehyde – its purity and stability, Histochemie *30*, 162.

Hayat, M. A. (1970), Principles and techniques of electron microscopy. Volume 1. Biological applications (Van Nostrand Reinhold, New York).

Hirsch, J. G. and M. A. Fedorko (1968), Ultrastructure of human leukocytes after simultaneous fixation with glutaraldehyde and osmium tetroxide and 'postfixation' in uranyl acetate, J. Cell Biol. *38*, 615.

Hökfelt, T. and G. Jonsson (1968), Studies on reaction and binding of monoamines after fixation and processing for electron microscopy with special reference to fixation with potassium permanganate, Histochemie *16*, 45.

Holt, S. J. and R. M. Hicks (1961), Studies on formalin fixation for electron microscopy and cytochemical staining purposes, J. biophys. biochem. Cytol. *11*, 31.

Hopwood, D. (1970), The reactions between formaldehyde, glutaraldehyde and osmium tetroxide, and their fixation effects on bovine serum albumin and on tissue blocks, Histochemie *24*, 150.

Ito, S. and M. J. Karnovsky (1968), Formaldehyde-glutaraldehyde fixatives containing trinitro compounds, J. Cell Biol. *39*, 168a.

Jacobs, G. F. and S. J. Liggett (1971), An oxidative-distillation procedure for reclaiming osmium tetroxide from used fixative solutions, Stain technol. *46*, 207.

Jard, S., J. Bourguet, N. Carasso and P. Favard (1966), Action de divers fixateurs sur la perméabilité et l'ultrastructure de la vessie de grenouille, J. Microscopie *5*, 31.

Karlsson, U. and R. L. Schultz (1965), Fixation of the central nervous system for electron microscopy by aldehyde perfusion. 1. Preservation with aldehyde perfusates versus direct perfusion with osmium tetroxide with special reference to membranes and the extracellular space, J. Ultrastruct. Res. *12*, 160.

Karnovsky, M. J. (1965), A formaldehyde–glutaraldehyde fixative of high osmolality for use in electron microscopy, J. Cell Biol. *27*, 137A.

Karnovsky, M. J. (1967), The ultrastructural basis of capillary permeability studied with peroxidase as a tracer, J. Cell Biol. *35*, 213.

Karnovsky, M. J. (1971), Use of ferrocyanide-reduced osmium tetroxide in electron microscopy, Proc. 14th Ann. Meeting Am. Soc. Cell Biol. p. 146.

Kaye, G. and M. J. Moses (1960), Combinations of formaldehyde and permanganate as fixatives for light and electron microscopy, Anat. Rec. *136*, 222.

Kelly, A. M. and S. I. Zacks (1969), The histogenesis of rat intercostal muscle, J. Cell Biol. *42*, 135.

Korn, A. H., S. H. Feairheller and E. M. Filachione (1972), Glutaraldehyde: nature of the reagent, J. Molec. Biol. *65*, 525.

Kushida, H. (1962), A study of cellular swelling and shrinkage during fixation, dehydration and embedding in various standard media, J. Electron Microscopy *11*, 135.

Lenard, J. and S. J. Singer (1968), Alteration of the conformation of proteins in red blood cell membranes and in solution by fixatives used in electron microscopy, J. Cell Biol. *37*, 117.

Lesseps, R. J. (1967), The removal by phospholipase C of a layer of lanthanum-staining material external to the cell membrane in embryonic chick cells, J. Cell Biol. *34*, 173.

Lewis, P. R., D. P. Knight and M. A. Williams (1974), Staining methods for thin sections, in: Practical methods in electron microscopy, A. M. Glauert, ed. (North-Holland, Amsterdam).

Luft, J. H. (1956), Permanganate – a new fixative for electron microscopy, J. biophys. biochem. Cytol. *2*, 799.

Luft, J. H. (1959), The use of acrolein as a fixative for light and electron microscopy, Anat. Rec. *133*, 305.

Luft, J. H. (1971), Ruthenium red and violet. 1 Chemistry, purification, methods of use for electron microscopy and mechanism of action, Anat. Rec. *171*, 347.

Malhotra, S. K. (1962), Experiments on fixation for electron microscopy. 1. Unbuffered osmium tetroxide, Q. Jl Microsc. Sci. *103*, 5.

Martin, R. and P. Rosenberg (1968), Fine structural alterations associated with venom action on squid giant axon fibers, J. Cell Biol. *36*, 341.

Maser, M. D., T. E. Powell and C. W. Philpott (1967), Relationships among pH, osmolality, and concentration of fixative solutions, Stain technol. *42*, 175.

Maunsbach, A. B. (1966a), The influence of different fixatives and fixation methods on the ultrastructure of rat kidney proximal tubule cells, I. Comparison of different perfusion fixation methods and of glutaraldehyde, formaldehyde and osmium tetroxide fixatives, J. Ultrastruct. Res. *15*, 242.

Maunsbach, A. B. (1966b), The influence of different fixatives and fixation methods on the ultrastructure of rat kidney proximal tubule cells. II. Effects of varying osmolality, ionic strength, buffer system and fixative concentration of glutaraldehyde solutions, J. Ultrastruct. Res. *15*, 283.

Millonig, G. (1961), Advantages of a phosphate buffer for OsO$_4$ solutions in fixation, J. appl. Phys. *32*, 1637.

Millonig, G. (1962), Further observations on a phosphate buffer for osmium solutions in fixation, Proc. 5th Int. Congr. Electron Microscopy, Philadelphia *2*, P-8.

Millonig, G. (1964), Study on the factors which influence preservation of fine structure, in: Symposium on electron microscopy, P. Buffa, ed. (Consiglio Nazionale delle Ricerche, Roma), p. 347.

Millonig, G. (1973), Personal communication, Laboratory of Molecular Embryology, Naples.

Millonig, G. and M. Bosco (1967), Some remarks on paraformaldehyde fixation, J. Cell Biol. *35*, 177A.

Millonig, G. and V. Marinozzi (1968), Fixation and embedding in electron microscopy, in: Advances in optical and electron microscopy, Vol. 2, R. Barer and V. E. Cosslett, eds. (Academic Press, New York), p. 251.

Mitchell, C. D. (1969), Preservation of the lipids of the human erythrocyte stroma during fixation and dehydration for electron microscopy, J. Cell Biol. *40*, 869.

Mollenhauer, H. H. (1959), Permanganate fixation of plant cells, J. biophys. biochem. Cytol. *6*, 431.

Moretz, R. C., C. K. Akers and D. F. Parsons (1969), Use of small angle X-ray diffraction to investigate disordering of membranes during preparation for electron microscopy. I. Osmium tetroxide and potassium permanganate, Biochim. biophys. Acta *193*, 1.

Napolitano, L. M., P. R. Sterzing and J. V. Scaletti (1969), Some observations on tissue fixed by glutaraldehyde–osmium tetroxide–digitonin mixtures, J. Cell Biol. *43*, 96a.

Palade, G. E. (1952), A study of fixation for electron microscopy, J. exp. Med. *95*, 285.

Palade, G. E. (1956), The fixation of tissues for electron microscopy, Proc. 3rd Int. Congr. Electron Microscopy, London, p. 129.

Palay, S. L., S. M. McGee-Russell, S. Gordon and M. A. Grillo (1962), Fixation of neural tissues for electron microscopy by perfusion with solutions of osmium tetroxide, J. Cell Biol. *12*, 385.

Pease, D. C. (1964), Histological techniques for electron microscopy, 2nd edition (Academic Press, New York and London).

Peracchia, C. and B. S. Mittler (1972), Fixation by means of glutaraldehyde–hydrogen peroxide reaction products, J. Cell Biol. *53*, 234.

Pollard, T. D. and S. Ito (1970), Cytoplasmic filaments of *Amoeba proteus*, J. Cell Biol. *46*, 267.

Porter, K. R. and R. D. Machado (1960), Studies on the endoplasmic reticulum. IV. Its form and distribution during mitosis in cells of the onion root tip, J. biophys. biochem. Cytol, *7*, 167.

Revel, J. P. and M. J. Karnovsky (1967), Hexagonal array of subunits in intercellular junctions of mouse heart and liver, J. Cell Biol. *33*, C7.

Rhodin, J. (1954), Correlation of ultrastructural organization and function in normal and experimentally changed proximal convoluted tubule cells of the mouse kidney, Thesis, Stockholm.

Riemersma, J. C. (1968), Osmium tetroxide fixation of lipids for electron microscopy. A possible reaction mechanism, Biochim. biophys. Acta *152*, 718.

Riemersma, J. C. (1970), Chemical effects of fixation on biological specimens, in: Some biological techniques in electron microscopy, D. F. Parsons, ed. (Academic Press, New York and London), p. 69.

Robertson, J. D. (1959), The ultrastructure of cell membranes and their derivatives, Biochem. Soc. Symp. *16*, 1.

Robertson, J. D., T. S. Bodenheimer and D. E. Stage (1963), The ultrastructure of Maunther cell synapses and nodes in goldfish brain, J. Cell Biol. *19*, 159.

Robison, W. G. and B. H. Lipton (1969), Advantages of dichromate–acrolein fixation for preservation of ultrastructural detail, J. Cell. Biol. *43*, 117a.

Rosenbluth, J. (1963), Contrast between osmium-fixed and permanganate-fixed toad spinal ganglia, J. Cell Biol. *16*, 143.

Ryter, A. and E. Kellenberger (1958), Étude au microscopie électronique de plasmas contenant de l'acide désoxyribonucleique, Z. Naturf. *13*, 597.

Sabatini, D. D., K. Bensch and R. J. Barrnett (1963), Cytochemistry and electron microscopy. The preservation of cellular ultrastructure and enzymatic activity by aldehyde fixation, J. Cell Biol. *17*, 19.

Sabatini, D. D., F. Miller and R. J. Barrnett (1964), Aldehyde fixation for morphological and enzyme histochemical studies with the electron microscope, J. Histochem. Cytochem. *12*, 57.

Sandborn, E. B., T. Makita and K.-N. Lin (1969), The use of dimethyl sulfoxide as an accelerator in the fixation of tissue for ultrastructural and cytochemical studies and in freeze etching of cells. Anat. Rec. *163*, 255.

Scallen, T. J. and S. E. Dietert (1969), The quantitative retention of cholesterol in mouse liver prepared for electron microscopy by fixation in a digitonin-containing aldehyde solution, J. Cell Biol. *40*, 802.

Schultz, R. L. and N. M. Case (1968), Microtubule loss with acrolein and bicarbonate-containing fixatives, J. Cell Biol. *38*, 633.

Schwab, D. W., A. H. Janney and J. Scala (1970), Preservation of fine structures in yeast by fixation in a dimethyl sulfoxide–acrolein–glutaraldehyde solution, Stain technol. *45*, 143.

Shea, S. M. (1972), Lanthanum staining of the surface coat of cells. Its enhancement by the use of fixatives containing alcian blue or cetylpyridinium chloride, J. Cell Biol. *51*, 611.

Shea, S. M. and M. J. Karnovsky (1969), The cell surface and intercellular junctions in liver as revealed by lanthanum staining after fixation with glutaraldehyde with added alcian blue, J. Cell Biol. *43*, 128a.

Silva, M. T., F. C. Guerra and M. M. Magalhães (1968), The fixative action of uranyl acetate in electron microscopy, Experientia *24*, 1074.

Silva, M. T., J. M. S.Mota, J. V. C. Melo and F. C. Guerra (1971), Uranyl salts as fixatives for electron microscopy. Study of the membrane ultrastructure and phospholipid loss in bacilli, Biochim. biophys. Acta *233*, 513.

Sjöstrand, F. S. and L.-G. Elfvin (1962), The layered asymmetric structure of the plasma membrane in the exocrine pancreas cells of the cat, J. Ultrastruct. Res. *7*, 504.

Smith, R. E. and M. G. Farquhar (1966), Lysosome function in the regulation of the secretory process in cells of the anterior pituitary gland, J. Cell Biol. *31*, 319.

Stefanini, M., C. De Martino and L. Zamboni (1967), Fixation of ejaculated spermatozoa for electron microscopy, Nature, Lond. *216*, 173.

Stein, O. and Y. Stein (1971), Light and electron microscopic radioautography of lipids: techniques and biological applications, Adv. Lipid Res. *9*, 1.

Strangeways, T. S. P. and R. G. Canti (1927), The living cell *in vitro* as shown by dark-ground illumination and the changes induced in such cells by fixing reagents, Q. Jl Microsc. Sci. *71*, 1.

Strauss, E. W. and A. A. Arabian (1969), Fixation of long-chain fatty acid in segments of jejunum from golden hamster, J. Cell Biol. *43*, 140a.

Tormey, J. McD. (1964), Differences in membrane configuration between osmium tetroxide – fixed and glutaraldehyde – fixed ciliary epithelium, J. Cell Biol. *23*, 658.

Tormey, J. McD. (1965), Artifactual localization of ferritin in the ciliary epithelium *in vitro*, J. Cell Biol. *25*, 1.

Trump, B. F. and R. E. Bulger (1966), New ultrastructural characteristics of cells fixed in a glutaraldehyde-osmium mixture, Lab. Invest. *15*, 368.

Trump, B. F. and J. L. E. Ericsson (1965), The effect of the fixative solution on the ultrastructure of cells and tissues, Lab. Invest. *14*, 1245.

Van Stevenick, M. E. (1972), Personal communication, University of Queensland.

Wetzel, B. K. (1961), Sodium permanganate fixation for electron microscopy, J. biophys. biochem. Cytol. *9*, 711.

Williams, N. E. and J. H. Luft (1968), Use of a nitrogen mustard derivative in fixation for electron microscopy and observations on ultrastructure of *Tetrahymena*, J. Ultrastruct. Res. *25*, 271.

Winborn, W. B. and L. L. Seelig (1970), Paraformaldehyde and *s*-collidine – a fixative for processing large blocks of tissue for electron microscopy, Tex. Rep. Biol. Med. *28*, 347.

Wood, R. L. and J. H. Luft (1965), The influence of buffer systems on fixation with osmium tetroxide, J. Ultrastruct. Res. *12*, 22.

Zamboni, L. and C. De Martino (1967), Buffered picric acid–formaldehyde: a new, rapid fixative for electron microscopy, J. Cell Biol. *35*, 148A.

Zetterqvist, H. (1956), The ultrastructural organization of the columnar absorbing cells of the mouse jejunum, Thesis, Stockholm.

# Chapter 3

# Fixation methods

Methods of fixing specimens for electron microscopy are as varied as the specimens themselves and only very general guide lines can be laid down. It is always advisable to first try a method that has been used successfully by other workers with similar specimens and then make modifications as required.

Biological specimens must be fixed as soon as possible after the death of the organism since alterations in fine structure occur rapidly (Fig. 3.1) (Trump et al. 1962). A few special tissues are unusually resistant to post-mortem changes (Ito 1962) but even for these it is advisable to fix the specimens without delay.

Many fixatives, including glutaraldehyde and osmium tetroxide, penetrate tissues slowly and in consequence the central region of a large specimen is rarely well fixed. The vehicle of the fixative usually penetrates more rapidly than the fixing agent itself and may cause damage before the fixation has effectively started. Palade (1956) has described a 'wave of acidification' which precedes the osmium during fixation with unbuffered osmium tetroxide. These damaging effects can be kept to a minimum by making the specimen as small as possible and by careful choice of the vehicle for the fixing agent (§ 2.2.1).

Even when the initial fixation is carried out *in vivo* (§ 3.3.2 and § 3.3.3.) it is usually necessary to complete the fixation by immersion of small pieces of tissue in the fixative. Since the specimens are small and some of the solutions used during fixation are expensive, it is convenient to use small disposable glass vials, about 2 cm in diameter and about 2 cm deep, with close-fitting plastic caps (Fig. 3.2.) for the fixation. In many procedures the specimens can remain in these vials throughout fixation, dehydration and

73

Fig. 3.1.   Transverse sections of rabbit cardiac muscle fixed (a) by perfusion, and (b) immediately after removal from the animal, in a mixture of paraformaldehyde and glutaraldehyde, followed by osmium tetroxide. Post-mortem changes are evident in (b), particularly the swelling of some mitochondria (arrow).

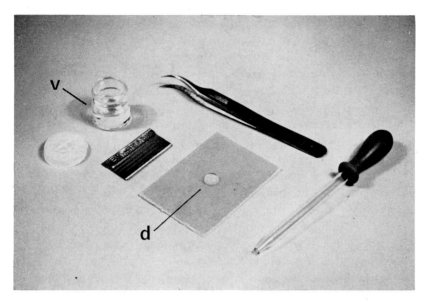

Fig. 3.2. The equipment required for fixation by immersion. Pieces of tissue are placed in a small drop of fixative (d) on a sheet of dental wax, and are then cut into small fragments with a sharp razor blade. These fragments are transferred to a small glass vial (v) containing fixative using a wide-mouthed pipette or pair of tweezers.

impregnation with the embedding medium, and need only be removed when placed in the capsules or moulds for the final embedding.

## 3.1 Time and temperature of fixation

### 3.1.1 PRIMARY FIXATION

The optimum time and temperature for primary fixation depends on the nature and size of the specimen and on the composition of the fixative. For isolated cells a few minutes fixation may be sufficient, while for compact tissues several hours may be required, particularly at low temperatures.

In general the time and temperature of fixation should be chosen so that the stabilising action of the fixative is complete, but fixation should not be continued beyond that point or excessive extraction by the buffer may occur. Fixation for 2 hr at 0 °C is common for primary fixation with osmium tetroxide, and similar times and temperatures are frequently used for primary aldehyde fixatives, in spite of the fact that they act more quickly, either as a result of their faster rate of reaction (e.g. glutaraldehyde), or faster rate of penetration (e.g. formaldehyde) or both (e.g. acrolein). Consequently, much shorter times could often be used with safety, and certainly the fixation should not be prolonged since aldehydes do not stabilise lipid components and these may well be extracted during an extended fixation.

Variation of the temperature of fixation with aldehydes or osmium tetroxide between 0 °C and 25 °C has little observable effect on fine structure, except for certain cell components, such as microtubules, which may not be preserved at low temperatures (e.g. see Gardner et al. 1969). At higher temperatures shrinkage of mitochondria and granularity of the cytoplasm have been reported (Rhodin 1954). Volume changes and extraction are reduced by lowering the temperature to 0–4 °C and this also has the advantage of slowing down enzymatic reactions and thus decreasing the extraction due to autolysis, although the rate of fixation is reduced at the same time. A short preliminary fixation in the cold, followed by the main fixation at room temperature has been recommended for osmium tetroxide (Pease 1964), glutaraldehyde and potassium permanganate (Millonig and Marinozzi 1968) fixation, to minimise extraction and to reduce the overall time of fixation.

### 3.1.2   WASH

When more than one fixation is being used the specimens are usually washed after the primary fixation to remove all unreacted fixative. It has been suggested that adequate washing is particularly important when aldehyde fixation is followed by osmium tetroxide fixation because the two fixatives react with each other (§ 2.8.2), but there is little published evidence of the adverse effects of a short wash, or none at all. For a few special specimens it is necessary to avoid washing; for example, Bodian (1970) found that storage for as little as 30 min in cacodylate buffer containing sucrose had a flattening effect on certain synaptic vesicles. Artefacts introduced during washing can be avoided by using a combined glutaraldehyde–osmium tetroxide fixative (§ 2.8.2).

It is usual, however, to use at least three changes of washing solution over a period of 2 hr. After glutaraldehyde fixation it is often convenient to store specimens in the washing solution overnight before further processing; this cannot be done after formaldehyde fixation because the reaction of formaldehyde with tissue components is reversible and extraction occurs (§ 2.5.1). For the majority of specimens the vehicle used in the primary fixative is a suitable washing solution, but sucrose may have to be added to adjust the osmolarity for osmotically-sensitive specimens. The washing solution may have to have a different osmolarity to the fixative because the osmotic properties of membranes often change during fixation (§ 2.2.3). Specimens are washed at the same temperature as the primary fixation unless they are to be stored for long periods when it is advisable to keep them in the cold.

### 3.1.3 SECONDARY FIXATION

The time and temperature for the second fixation are much less critical than for the first. Most commonly specimens are fixed in 1% osmium tetroxide in the same buffer used for the primary fixation for 1 hr at 4 °C or at room temperature.

## 3.2 Fixation schedules

The sequence of fixation with an aldehyde, followed by osmium tetroxide and possibly also uranyl acetate, has become the standard fixation schedule at the present time for a great variety of specimens.

A representative fixation schedule is:

1. *First fixation* for 1 hr at 4 °C in 2% paraformaldehyde plus 2.5% glutaraldehyde in 0.1 M phosphate buffer, pH 7.4, containing 2.5 mM calcium chloride.
2. *Wash* for 2 hr (or overnight) at 4 °C in 3 changes of 0.1 M phosphate buffer, pH 7.4, containing 2.5 mM calcium chloride.
3. *Second fixation* for 1 hr at room temperature in 1% osmium tetroxide in 0.1 M phosphate buffer, pH 7.4, containing 2.5 mM calcium chloride.

Alternatively, when specimens are to be fixed or 'block stained' with uranyl acetate a different schedule is used because uranyl ions form precipitates in phosphate and cacodylate buffers.

1. *First fixation* for 1 hr at 4 °C in 2% paraformaldehyde plus 2.5% glutaraldehyde in 0.1 M phosphate buffer, pH 7.4, containing 2.5 mM calcium chloride.
2. *Wash* for 2 hr (or overnight) at 4 °C in 3 changes of 0.1 M phosphate buffer, pH 7.4, containing 2.5 mM calcium chloride.
3. *Second fixation* for 1 hr at room temperature in 1% osmium tetroxide in veronal–acetate buffer, pH 7.4, containing 2.5 mM calcium chloride.
4. *Rinse* in veronal–acetate buffer, pH 7.4.
5. *Third fixation* for 1 hr at room temperature in 0.5% uranyl acetate in veronal–acetate buffer, pH 5.0.

## 3.3   Organs and tissues

### 3.3.1   IMMERSION FIXATION

Organs and tissues, such as the skin, that can tolerate a temporary interruption to their blood supply and still retain their function and structure,
can be fixed by immersion *in vitro*. The animal is killed and then the tissue
is removed by dissection as rapidly as possible. A relatively large piece of
tissue is taken and then cut into smaller pieces for fixation, the size of the
piece depending on the nature of the fixative and the density of the tissue.
For primary fixation with osmium tetroxide the tissue is cut into small
cubes which must not be larger than 0.5 mm on each side if the whole cube
is to be well fixed, while for aldehyde fixatives the pieces of tissue can be
considerably larger. Small cubes are then cut later from selected regions of
these pieces while they are in the buffer wash before osmium tetroxide
fixation. The proportion of the larger pieces that is well fixed depends upon
the aldehyde fixative; for glutaraldehyde, which penetrates slowly, the zone
of adequate fixation may well be only 0.5 mm deep, while for a fixative
containing formaldehyde this zone will be considerably deeper, but will
probably not be more than a few millimetres (Hayat 1970). In general, the
small cubes should be cut from as near the surface of the piece of tissue as
possible, while avoiding the surface itself which will have been damaged
when the piece was cut out. The fact that the cubes can be selected in this
way is a great advantage of the double fixation with an aldehyde followed
by osmium tetroxide, as compared with osmium tetroxide alone. A relatively
large proportion of a small cube cut from the fresh tissue for osmium
tetroxide fixation is damaged during cutting, while after aldehyde fixation
the tissue is much tougher and the mechanical damage produced during
cutting into cubes is considerably less.

   A survey of the recent literature indicates that the most popular primary
fixatives for immersion fixation of a wide range of specimens are either
1.7 to 4% glutaraldehyde in 0.1 M phosphate or cacodylate buffer, pH
7.2–7.4 (§ 2.4.3), or diluted Karnovsky's (1965) fixative, consisting of 2%
paraformaldehyde, 2.5% glutaraldehyde and 2.5 mM calcium chloride in a
similar buffer (§ 2.5.5).

   The cutting of the tissues into pieces or cubes is conveniently done by
placing the tissue on a sheet of dental wax in a fume cupboard, adding
enough fixative to keep the tissue moist, and then cutting the tissue cleanly
with a new, single-edged razor blade or dissecting knives (Fig. 3.2). Great
care is required during cutting to ensure that the tissue is damaged as little

as possible. As soon as the pieces or cubes are prepared they are transferred to pre-labelled glass vials containing the fixative (Fig. 3.2). The partially-fixed tissue is easily damaged and is best transferred using a wide-mouthed Pasteur pipette or pieces of filter paper.

The relative volumes of the specimen and the fixative are important; if there is too little fixative the effective concentration of the fixative is reduced by dilution with soluble components of the specimen. In general, the fixative should be at least 10 times greater in volume than the specimen. If fixation is carried out at 0–4 °C the vials containing the fixative are placed in a bath of crushed ice some time before fixation is due to start. Specimens are fixed for 30 min to 1 hr at room temperature or for 1 to 4 hr at 0–4 °C.

After primary fixation the fixative is removed with a fine pipette and flushed down the sink in the fume cupboard with a large quantity of water. The fixative is replaced with the first washing solution (§ 3.2) which is usually the same buffer as used for the fixative with the addition of sucrose, if necessary. The pieces of tissue are then removed from the vials, cut into small cubes not greater than 0.5 mm on each side, and replaced in the vials with fresh washing solution. At this stage the fixed tissue is usually tough enough to be handled with a fine pair of forceps, although great care should be taken at all stages to be as gentle as possible. The chosen fixation schedule is followed (§ 3.2), the specimens remaining in the glass vials throughout the procedure. When fixation or washing is being done in the cold, the vials are transferred to a refrigerator, but all changes of fixative must be done in a fume cupboard.

Marks and Briarty (1970) describe the use of a disposable syringe fitted with a filter as an alternative to a small vial for processing specimens through fixatives and dehydrating agents.

### 3.3.2 *IN VIVO* FIXATION

The fine structure of many tissues is critically dependent on a continuous blood supply and it is essential that fixation of such tissues should be started while the animal is still alive. For fixation *in vivo* the animal is anaesthetised, the tissue exposed by dissection, and then the fixative is applied by dripping it over the surface of the organ or by injecting it into the tissue.

The dripping method is only useful when the surface layers of an organ are being studied since the fixative does not penetrate through the surface capsule and any attempt to remove the capsule before fixation starts may damage the structure of the organ (Maunsbach et al. 1962). For fixation by this method the cold fixative is dripped onto the surface of the organ continuously for 10 to 20 min, the excess fixative being collected with a suitably placed pad of cotton wool. The method has mainly been used with veronal–acetate buffered osmium tetroxide fixatives (§ 2.3.5) at concentrations of 1–2%. Since it is limited to the study of surface layers of organs it has largely been supplanted by techniques in which the fixative is injected into the tissue, or by fixation by perfusion (§ 3.3.3).

For the examination of the internal structure of organs the fixative is injected into the tissue using a micropipette or hypodermic needle. The fixative is injected slowly for 2 to 15 min and is usually cooled to 0–4 °C. Some typical references are listed in Table 3.1.

After initial fixation *in vivo*, by dripping or injection, the organ is removed and cut into small pieces and the primary fixation completed by immersion in the fixative in the cold or at room temperature (see § 3.3.1), the total fixation time varying from 1 to 4 hr.

These methods of fixation, in common with perfusion fixation, have the disadvantage that the anaesthetic may have an effect on the fine structure of the tissue under study; these effects are reduced to a minimum by starting the fixation as soon as possible after the administration of the anaesthetic. The possible adverse effects of anaesthetics can be avoided by killing the animal by a blow on the head, but it is then necessary to work extremely rapidly since the heart will not continue to beat for long.

### 3.3.3  PERFUSION FIXATION

The method of *in vivo* fixation described in the previous section (§ 3.3.2) is usually adequate for small organs, or when the surface regions of larger organs are being studied, but poor fixation is obtained of internal regions of organs such as the kidney and the lung, which have a copious blood supply, and of the brain and other tissues of the central nervous system. For these tissues it is necessary to perfuse the fixative through the living, anaesthetised animal.

There are many methods of perfusion recorded in the literature, and probably many more unpublished variants peculiar to a particular laboratory. For convenience of description perfusion procedures will be discussed under three headings: the type of anaesthesia, the route by which the fluids are injected, and the composition of the perfusion fluids.

### 3.3.3.a.  *Anaesthesia*
The type of anaesthetic used is usually determined by the preferences of the investigator and is likely to be important only in specific physiological

TABLE 3.1

*In vivo fixation by injection*

| Organ | Site of injection | Fixing agent | Vehicle | pH | Temp. | Time | Reference |
|---|---|---|---|---|---|---|---|
| Bullfrog heart | Sinus | 0.88% G | 0.067 M cac | 7.2 | 0°C | ? | Baldwin (1970) |
| Mouse diaphragm | Pleural and abdominal cavities | 2% P + 2% G | 0.09 M cac + 0.025% $CaCl_2$ | 7.4 | r.t. | 3–5 hr | Karnovsky (1967) |
| Rat pituitary | Cranial cavity | (a) 1% $OsO_4$ | phosphate | 7.6 | ? | few min | Smith and Farquhar (1966) |
| Rat pituitary | Cranial cavity | (b) 1.5% G | 0.067 M cac + 1% sucrose | 7.4 | ? | few min | |
| Rabbit cartilage | Knee joint | 3% G | phosphate + $CaCl_2$ | 7.3 | ? | few min | Palfrey and Davies (1966) |
| Rat kidney | Lumen of tubule | 3% G | 0.1 M cac | 7.2 | 0°C | 2 min | Maunsbach (1966) |
| Goldfish brain | Ventricle of heart | 1.5% $KMnO_4$ | 1.5% $NaHCO_3$ + salts | 7.4 | r.t. | approx. 10 min | Robertson (1963) |
| Bat stomach | Stomach | (a) 1% $OsO_4$ | VeAc | 7.1–7.5 | cold | few min | Ito and Winchester (1963) |
| Bat stomach | Stomach | (b) 1.33% $OsO_4$ | collidine + 0.25 M sucrose | 7.1–7.5 | cold | few min | |
| Rabbit eye | Posterior chamber | 1% $OsO_4$ | VeAc + salts | 7.2–7.6 | cool | 15 min | Tormey (1963) |
| Cat pancreas | Pancreas | (a) 1% $OsO_4$ | VeAc | 7.2–7.4 | 2–5°C | 10–15 min | Sjöstrand and Elfvin (1962) |
| Cat pancreas | Pancreas | (b) 0.6% $KMnO_4$ | VeAc | 7.2–7.4 | 2–5°C | 10–15 min | |

*Key*: G, glutaraldehyde; P, paraformaldehyde; VeAc, veronal-acetate; r.t., room temperature; cac, cacodylate.

or pharmacological studies. Ether and halothane are the most commonly used of the volatile anaesthetics and are particularly convenient for the smaller laboratory animals, such as rats, mice and guinea pigs. For larger animals it is usually more convenient to give an intra-peritoneal injection, which is easy, or an intravenous injection, which is more difficult, and to use nembutal, chloral hydrate or urethane. Until the operator has gained some skill at perfusion a very deep level of anaesthesia should be aimed at; administration of an overdose should do no harm. The dose of anaesthetic to use can be found by consulting a standard laboratory manual (such as Lumb 1963).

### 3.3.3b.   *The route of injection of perfusion fluids*

The route by which the perfusion fluid is injected is determined largely by the size of the animal, the tissue to be perfused and the surgical skill of the operator. The commonest method of controlling the rate of perfusion is to use a simple gravity feed, but for small laboratory animals it is often sufficient to use a manually operated injection syringe which can be safely refilled several times during the perfusion, provided that care is taken not to dislodge the perfusion needle or to introduce air bubbles. For more sophisticated experiments a perfusion pump can be used (e.g. Gil and Weibel 1969–70).

The perfusion pressure must be high enough to ensure good flow of the fixative but not significantly higher than the arterial pressure in the intact animal. With an intra-arterial gravity feed, the level of the reservoir of perfusion fluid should be 120–150 cm (4–5 ft) above the animal, but when an intravenous route is used, as in perfusion of the liver via the portal vein, the fluid level should be only 20–30 cm above the animal. Animals should always be perfused in a good fume cupboard or under an extraction hood, and the operator must wear protective goggles.

For small animals the most satisfactory route for perfusion is through the left ventricle into the base of the aorta (see below). For young animals a more practicable method is simply to insert a fine hypodermic needle connected to a reservoir of perfusion fluid through the wall of the left ventricle and hold it in place while fixation occurs. As soon as the needle is in place the right atrium should be cut with a fine pair of scissors to allow venous blood and perfusate to escape.

For perfusion of mature rats the following procedure is recommended:
1.   When the animal is deeply anaesthetised, lay it on its back in a shallow tray.

2. Make a midline incision down the whole length of the thorax and much of the abdomen and reflect the skin.
3. Expose the heart by cutting the ribs and intercostal muscles along each side and raising the flap so formed.
4. Clean away the thymus and excess adipose tissue from the arch of the aorta and draw a ligature beneath it ready to tie the perfusion cannula in place.
5. Make a stitch through the apex of the heart and clip the thread to the side of the animal so that the heart is under slight tension.
6. Make a small perforation in the wall of the fourth ventricle with fine scissors and quickly insert a cannula into the opening and through the aortic valve so that the cannula projects about 5 mm into the aorta.
7. Tie the cannula in place and as soon as the perfusion fluid is flowing cut the right atrium.

For mice it may prove more convenient to use a cannula made out of plastic tubing. Force a short length of tubing just small enough to enter the aorta into wider tubing so that about 3 mm is projecting. When the tubing is passed through the ventricle, the wider tubing will not enter the aortic valve and the fine tubing will project just far enough to provide an adequate seal.

The method of perfusion described above is very wasteful of fixative in large animals and, if tissues are only required from the head end, the abdominal aorta should be clamped off as high up as possible once the perfusion has begun. For some purposes, in fact, it may be more satisfactory to perfuse via the abdominal aorta in animals larger than the rat, and it is essential to use this route in order to perfuse the heart properly. When perfusing through the aorta it is advisable to clip the aorta in two places and then insert the needle between. To release blood it is best to cut the jugular vein and then the inferior cava.

### 3.3.3c  *Fluids and fixatives for perfusion*
There is no general agreement in the literature about the optimum temperature of the perfusion fluid, or whether the blood should be washed out with saline before the fixative is introduced. For preliminary experiments it is sufficient to use fixative alone at room temperature. If saline is used, a convenient way of ensuring that some saline precedes the fixative in perfusing a small animal is to fill the tubing with saline before carefully filling the reservoir with fixative. For larger animals it is necessary to have

two reservoirs, one with saline and the other with fixative, connected to the perfusion tubing via a three-way tap.

For small animals, where the perfusion rate is low, the temperature of the perfusion fluid can be adequately controlled by passing it through a coil immersed in a water bath. For larger animals the fluid in the reservoir must be kept at an appropriate temperature. This is done most easily by insulating the reservoir with a jacket of expanded polystyrene or similar material.

Fixation is undoubtedly more rapid and effective the higher the temperature. For purely morphological studies, therefore, the perfusion fluid should probably be near the deep body temperature of the animal. For enzyme histochemistry the perfusion fluid should be at room temperature, and for very labile enzymes it may be necessary to cool the fixative down to near 0 °C (Lewis et al. 1974). Perfusion fluids below body temperature are liable to cause vasoconstriction which can impair the effectiveness of the perfusion.

Opinions differ on the optimum concentration of the fixing agent in the initial perfusing fluid. Peters et al. (1968) and Brightman and Reese (1969) followed the lead of Reese and Karnovsky (1967) in using a small volume of a dilute aldehyde fixative first. In contrast, Schultz and Case (1970) specifically blame swollen mitochondria on slow initial fixation and they advocate the use of a small volume of very concentrated fixative (e.g. 19% glutaraldehyde) to precede the main perfusion fluid. With a concentrated initial fixative, a small volume of a saline solution must be used first to wash out the blood. Again a compromise may be necessary; when tissue is to be used for enzyme histochemistry a dilute fixative may have to be used, while for purely morphological studies rapid initial fixation is probably best.

Bohmann and Maunsbach (1970) studied the effect of adding dextran or polyvinylpyrrolidone (PVP) to the fixative to raise the colloid osmotic pressure and reported much less expansion of extravascular spaces. The dextran and PVP which they used had molecular weights of about 40,000 daltons and were at a final concentration of 2–4% in the perfusion fluid. Schultz and Case (1970) used 0.5% or 1% gum acacia in their perfusion fluids; they were troubled by turbidity when gum acacia was present and cured it by the addition of a small quantity of polyphosphate. It seems likely, however, that the turbidity was caused by the very high concentration of phosphates in their perfusion fluid.

For purely morphological studies the following procedure should be suitable for preliminary experiments. It is a simplified modification of the method described by Schultz and Case (1970) and the volumes quoted are for perfusion of an adult rat through a cannula inserted via the left ventricle into the root of the aorta.

1. Begin the perfusion with 10 ml of an isotonic buffered saline solution containing 2% PVP or dextran (of molecular weight in the range 40,000 to 50,000 daltons) or 1% gum accacia. For critical experiments a complete Tyrode solution should be used, but for many purposes some of the minor constituents can be omitted.

2. Continue the perfusion with 15–18 ml of a concentrated glutaraldehyde solution containing:

    15–20% glutaraldehyde
    2% PVP or dextran, or 1% gum acacia
    20–50 mM of a suitable buffer at pH 6.8–7.4
    physiological saline to give an osmolarity of 300 mosmols without the aldehyde

    An example would be a solution prepared by dissolving 0.5 g PVP in 5 ml 0.2 M sodium cacodylate plus 2.5 ml 1.0 M NaCl and 0.5 ml 0.2 M calcium acetate or chloride, adding 10 ml of 50% glutaraldehyde, adjusting the pH with HCl if necessary and making up to a total of 25 ml with distilled water. The solution should be filtered and must be crystal clear before use.

3. Complete the perfusion with 200–250 ml of the main fixative, which is usually a more dilute glutaraldehyde solution (e.g. made up in the same way as the concentrated fixative but with 4 or 5% glutaraldehyde instead of 20%) or a combined paraformaldehyde-glutaraldehyde solution (e.g. Reese and Karnovsky 1967). The ionic composition of the main fixative should not differ significantly from the solution used in step 2.

4. Dissect out the tissues required and continue the fixation by immersion (§ 3.3.1) either in the solution used in step 3, or in an equivalent solution containing only paraformaldehyde, at 4 °C for 2–4 hr.

For enzyme cytochemistry it is advisable to omit step 2 and perhaps also to replace the saline of step 1 with a very dilute fixative about one fifth the strength of the main perfusion fluid (Reese and Karnovsky 1967). For some enzymes perfusion must be discontinued after only 2–3 min to prevent excessive inactivation (Lewis et al. 1974).

3.3.4 FIXATION OF VERY LARGE SPECIMENS

It is sometimes necessary, as in the study of surgical biopsies, to fix very large specimens. A primary fixative is required which has the maximum penetrating power, so that aldehyde fixatives containing acrolein or para-formaldehyde are chosen and trinitro compounds added (§ 2.8.4). A suitable fixative is one containing 2.5% glutaraldehyde and 2% paraformaldehyde in 0.1 M cacodylate buffer, pH 7.4, containing 0.2% trinitrocresol and 2 mM calcium chloride.

Alternatively, a rapidly penetrating fixing agent (e.g. paraformaldehyde) at high concentration (10%) is buffered with collidine (0.2 M) (§ 2.2.8) which extracts tissue components and aids penetration. Winborn and Seelig (1970) obtained good fixation in the centres of very large tissue blocks (3.5 cm$^3$) with such a fixative.

### 3.3.5 ORGAN CULTURES AND COLONIES GROWING ON AGAR

Organ cultures are very easy to fix since all that is necessary is to lift them from the medium in which they have been grown, rinse them in a balanced salt solution to remove excess medium and place them directly in the primary fixative in a small vial. They are then handled in exactly the same way as pieces of tissues fixed by immersion (§ 3.3.1), using a similar fixation schedule (§ 3.2).

Colonies of cells (tissue culture cells, bacteria, fungi, etc.) growing on agar or other solid medium are fixed by the same method (e.g. Glauert and Hopwood 1960; Tuffery 1972).

## 3.4  Botanical specimens

In comparison with animal tissues, botanical tissues are often believed to be relatively unaffected by removal from the plant, but ultrastructural changes are known to occur quite rapidly, mainly due to water loss. Pieces of leaves, stems or roots are therefore fixed immediately after removal from the plant. A drop of fixative is added to the required sample which is then cut, with a sharp razor blade, into thin slices (0.2–1.0 mm thick) or small cubes (about 1.0 mm$^3$). These pieces are then immersed in fresh fixative as quickly as possible, taking care to minimise structural damage at this stage. It is some-times advantageous to expose tissues which float in the fixative to reduced pressure for a brief period to remove internal air and thus ensure complete contact with the fixative solution.

The early osmium tetroxide fixatives used for animal tissues gave poor results when applied to plant specimens, probably because the vacuolar contents are released at that stage in fixation when the surrounding membrane becomes permeable. The first successful preservation of plant cells was achieved using potassium permanganate (0.6–5.0%), either unbuffered or buffered with veronal–acetate, for $\frac{1}{2}$–1 hr at 2–4 °C or at 22 °C (Mollenhauer 1959). Fixation with permanganate gives excellent visualization of cellular membranes, but also causes the progressive removal of other cellular components (§ 2.7.1), and should therefore be used only for special purposes, e.g. when studying the extent of cellular membrane systems such as the endoplasmic reticulum.

Since the introduction of aldehyde fixatives, much improved preservation of cytoplasmic detail is obtained using a preliminary fixation with an aldehyde (frequently 2–6% glutaraldehyde in phosphate or cacodylate buffer, see Table 3.2), followed, after thorough rinsing, by fixation with osmium tetroxide (usually 1–2% in the same buffer). The duration of both the aldehyde and osmium tetroxide fixations may be varied and may be carried out at room temperature or in the cold. For example, a common procedure is: fixation in glutaraldehyde for $1\frac{1}{2}$–4 hr at room temperature, followed by osmium tetroxide fixation for 1–2 hr at room temperature. For some specimens the duration of fixation can be shortened considerably with equally good results (Bain and Gove 1971).

Apart from variations in the duration and temperature of fixation, there are many other procedures which are followed by different workers, such as the addition of salts or sucrose to fixatives to increase the osmolarity, or the use of mixtures of fixatives to give improved preservation (Table 3.2).

Most specimens are fixed by the simple immersion method already described, but modifications are required depending on the nature of the specimen, its size and the ease of penetration of the fixative. Small specimens, such as algal unicells or isolated chloroplasts, are fixed as a centrifuged pellet (§ 3.6.1) or in suspension (§ 3.6.2). Larger specimens, such as pieces of stem or tissues containing glands, are given a brief preliminary fixation before further dissection to obtain the required groups of cells. It is sometimes advisary to start the fixation before the tissue has been completely removed from the plant. For example, Robards (1968) irrigated the stem with cold fixative during excision in a study of differentiating xylems of willow. Some of the variations in fixatives and methods of application which have been used successfully in the study of particular specimens are indicated in Table 3.2. Since the suitability of a fixation technique can be determined only by trial and error, it is frequently advisable to compare the results of two or more procedures.

TABLE 3.2

Fixation of botanical specimens

| Specimen | Fixation method | Fixing Agent | Vehicle | pH | temp. | time | Reference |
|---|---|---|---|---|---|---|---|
| Wheat coleoptiles and root tips | Immersion | 6% G | 0.02 M ph + CaCl₂ | 7.0 | r.t. | 1–2 hr | Pickett-Heaps and Northcote (1966) |
| Pinus cambial derivatives | Immersion | (a) 3% G | 0.025 M ph | 6.8–7.0 | r.t. then 5°C | 2 hr ⎫ 22 hr ⎬ | Srivastava and O'Brien (1966) |
|  |  | (b) 10% A | 0.025 M ph | 6.8–7.0 | 0°C | 6 hr ⎭ |  |
| Phalaris root tips | Immersion | 2% KMnO₄ | Unbuffered | ? | 20–22°C | 2 hr | López-Sáez et al. (1966) |
| Hibiscus pods | Immersion | 3% A + 5% G | cac | ? | cold | 1 hr | Mollenhauer (1967) |
| Capsella embryos | Immersion | (a) 2% KMnO₄ | Unbuffered |  | 4°C | 15–23 hr ⎫ | Schulz and Jenzen (1968) |
|  |  | (b) 6% G | 0.06 M ph | 6.8 | 4°C | 4 hr ⎬ |  |
| Venus flytrap digestive gland | Immersion | (a) 2.5% G | 0.02 M coll | 7.4 | 3°C | 24 hr ⎫ | Scala et al. (1968) |
|  |  | (b) 1% OsO₄ | VeAc + 0.015 M sucrose | 7.4 | 3°C | 2.5 hr ⎬ |  |
|  |  | (c) 2.5% G and then 3% KMnO₄ | 0.02 M coll | 7.4 | 3°C | 2 hr ⎬ |  |
|  |  |  | Unbuffered |  | ? | ? ⎭ |  |
| Willow xylem | Irrigation and then immersion | (a) 6% G | 0.05 M cac + 0.15 M sucrose | 7.0 | 4°C | 18 hr ⎫ | Robards (1968) |
|  |  | (b) 3% G | 0.05 M cac + 0.15 M sucrose + salts | 7.2 | r.t. then 4°C | 1 hr ⎬ 4 hr ⎭ |  |

| | | | | | | | |
|---|---|---|---|---|---|---|---|
| *Myosurus* female gametophyte | Immersion | (a) 1% KMnO₄ (b) 6% G | Unbuffered Phosphate | 7.2 | r.t. 4°C | 2 hr 24 hr | Woodcock and Bell (1968) |
| Bean leaves | Immersion | 2% G + 2% OsO₄ | 0.05 M cac | 7.0 | 0–2°C | ¼ hr | Falk (1969) |
| Tobacco pith culture | Immersion | 4% P + 5% G | 0.1 M ph | 6.8 | ? | 2 hr | Anderson and Cronshaw (1970) |
| Brown alga | ? | 2.5% G + 2.5% P | Cac in sea water | 6.8 | r.t. | 1–2 hr | Bisalputra and Bisalputra (1969) |
| Spinach isolated chloroplasts | Suspension then pellet | 1% G (final conc.) 5% G | 0.05 M ph | 6.8–7.0 | ? | 2 hr | Nobel et al. (1966) |
| *Chlorella* autospores | Suspension | 0.5% G + 0.4% P then 2.5% G | Diluted culture medium Diluted culture medium | | | 1 hr 1 hr | Atkinson et al. (1972) |
| Fungus zoospores | Suspension then Millipore filter | 1.5% G (final conc.) 1.5% G | 0.005 M ph ph | 6.8 6.8 | 22°C 22°C | 10 min 1.5 hr | Hoch and Mitchell (1972) |
| Rice embryos | In nylon sacs weighted with glass tubing | 1–6% G | 0.05 M ph | 7.2 | 4°C | 12 hr | Öpik (1972) |

*Key:* G, glutaraldehyde; A, acrolein; P, paraformaldehyde; ph, phosphate; cac, cacodylate; coll, *s*-collidine; r.t., room temperature.
*Note.* Aldehyde fixation is followed by fixation with osmium tetroxide.

## 3.5  Monolayers of cells

Cells in culture growing as a monolayer on glass, plastic or other substrate, such as a Millipore filter, are easily fixed by immersing the whole culture on its support in the fixative. It is convenient to use Columbia staining jars in which the coverslips or other supports can stand vertically and in which a number of cultures can be fixed at the same time. Alternatively, the culture is placed face-upwards in a Petri dish and is covered with a shallow drop of the fixative, care being taken to use sufficient fixative, or a drop of fixative is placed in the bottom of a Petri dish and the culture placed face-downwards on top of it. The lid is placed on the Petri dish so that the fixative does not dry out during the period of fixation and the dish is placed on a slow shaker (§ 5.2), to assist the circulation of the fixative.

Monolayers grown directly in plastic Petri dishes, or other containers, are fixed by replacing the medium with sufficient fixative to completely cover the layer of cells; about 5 ml of fixative is required for a dish 8 cm in diameter.

When studying cell surfaces it is advisable to rinse the monolayer briefly (e.g. 1 min at 37 °C) in a balanced salt solution before fixation to remove all traces of the culture medium. Otherwise components of the medium will be fixed to the surfaces of the cells.

Although this method of fixation is simple, care has to be taken in the choice of fixative and buffer since the cells are relatively unprotected against osmotic changes, as compared with cells in tissues (§ 2.2). A survey of the recent literature indicates that satisfactory preservation is usually obtained by primary fixation for 15 to 30 min at room temperature in 1–3% glutaraldehyde in 0.1 M phosphate or cacodylate buffer, pH 7.2–7.4, containing 2–3 mM calcium chloride (§ 2.4.3). Some workers prefer to use Karnovsky's (1965) paraformaldehyde–glutaraldehyde fixative (§ 2.5.5), diluted 1:2 or 1:4 with buffer. If the fixation is not satisfactory it is worthwhile testing the mixed glutaraldehyde-osmium tetroxide fixative of Hirsch and Fedorko (1968) (§ 2.8.2), fixing the monolayer for 15 to 30 min at 4 °C. In studies in which the cell surface is not being examined in detail, good fixation is obtained by fixing the cells in the presence of the growth medium. The medium is not removed and the culture is fixed by adding a volume of warm (37 °C) double-strength fixative equal in volume to the growth medium (Ross 1972).

As an alternative to immersion fixation, the monolayer can be fixed by exposing the cells to the vapours of the fixative while they are bathed in the culture medium or in a physiological salt solution (e.g. see Porter and Kall-man 1953). The monolayer on its support and an open vial containing the fixative are placed together in a closed container, such as Petri dish, and left for a few minutes at room temperature.

After primary immersion or vapour fixation with an aldehyde fixative, the monolayer is rinsed in a washing solution and then fixed for 15 to 30 min at room temperature in 1% osmium tetroxide in the same buffer. If the monolayer is subsequently fixed with uranyl acetate, a period of 15 to 30 min is again sufficient. Thus the schedule for monolayers is similar to that for pieces of tissue (§ 3.2) but the time in each solution is considerably reduced.

Special problems arise during the embedding procedure, particularly when the layer of cells has to be removed from its support. Techniques for embedding monolayers are described in § 5.8.2.

## 3.6 Isolated cells

The fixation methods for isolated cells (cells in culture, protozoa, bacteria, blood cells etc.) are very different from the methods used for tissues. There is now no problem in obtaining rapid fixation, but it is sometimes difficult to handle the cells, particularly if only small quantities are available, and, as for monolayers of cells, the composition of the fixative has to be chosen with care.

### 3.6.1 FIXATION IN A PELLET

It is often possible to obtain good fixation of mammalian cells by centri-fuging the cells into a pellet, carefully removing the supernatant without disturbing the pellet, and then adding the fixative. The fixative is run slowly down the side of the centrifuge tube so as to keep the pellet intact; the volume of fixative added should be at least ten times the volume of the pellet. For larger pellets, more than a few mm in depth, the cells should be resuspended in the fixative by gentle shaking to ensure that all the cells are adequately fixed. After the primary fixation the cells are then centrifuged to a pellet again before the next step in the preparative procedure. One of the fixatives used for monolayers of cells (§ 3.5) is suitable.

After the primary fixation is complete the fixative is carefully removed from the pellet and replaced with the washing solution. The cohesion of the fixed pellet depends on the type of cell, the composition of the fixative and the speed of centrifugation. Many pellets which have been fixed in glutaraldehyde are firm and show no tendency to disintegrate in the washing solution, and these can be removed from the centrifuge tube at this stage. The fact that the pellet has to be removed in this way should be borne in mind when selecting the centrifuge tube (3 ml conical tubes are suitable for processing most cells). If glass tubes are used a special set of tubes should be kept for the purpose since it is difficult to remove all traces of fixative from them. The pellet can usually be freed from the bottom of the tube fairly easily using the end of a fine spatula. If the pellet is difficult to dislodge it is preferable to use plastic tubes which can be cut down to remove the pellet.

When only a small quantity of cells is available the method developed by Malamed (1963) for cell fractions (§ 3.8) is useful (Farquhar et al. 1972).

After removal from the centrifuge tube the pellet is placed in a small drop of washing solution on a sheet of dental wax and cut into small cubes by the same method used for pieces of tissue (§ 3.3.1). Care should be taken to cut out cubes from different regions of the pellet so as to check the homogeneity of the original suspension of cells. These cubes are then passed through the subsequent steps of the preparative procedure, the fixation schedule being similar to that used for immersion fixation of cubes of tissue (§ 3.2); a second fixation for 1 hr at room temperature or 4 °C in an osmium tetroxide fixative is customary.

Pellets of fixed cells which tend to disintegrate are encapsulated by one of the methods described in § 3.6.3.

Anderson (1965) and Watanabe et al. (1967) used a modification of the method of fixing cells in a pellet as a simple technique for preparing peripheral leucocytes for electron microscopy. Anticoagulated blood is centrifuged in a horizontal type head and all but the last drop of plasma removed with a Pasteur pipette without disturbing the buffy coat or tilting the tube. A glutaraldehyde fixative is then carefully layered over the buffy coat with a pipette, using approximately the same volume of fixative as the removed plasma, and the tube is left in the vertical position for 15 min or more. The disc of fixed leucocytes, embedded in solidified plasma, can then be removed with only a thin layer of adhering erythrocytes. This disc is then placed in fresh fixative for a further period, if required, and then rinsed in several changes of buffer and cut into small cubes for further processing, care being taken to preserve the orientation of the layers.

Successful results have been obtained with this method with a variety of aldehyde fixatives; Anderson (1965) and Watanabe et al. (1967) used glutaraldehyde fixatives ranging in concentration from 2 to 6.25%, in 0.1 M phosphate or cacodylate buffer, pH 7.4, at room temperature (for 15 min) or at 4 °C (for 1 hr), while Carson et al. (1972) tested a range of paraformaldehyde fixatives of different concentrations and in different buffers, at pH 7.32 to 7.45, and found that phosphate buffers gave the most reliable and uniform fixation (Fig. 2.3). Variations in the concentration of paraformaldehyde from 0.5 to 4% and in temperature (4 or 25 °C) had little noticeable effect. All specimens were fixed for a minimum of 18 hr. The discs of blood cells were very difficult to handle because they are not cohesive after paraformaldehyde fixation.

Carson et al. (1972) also describe two other, more complicated, fixation methods for blood cells, and a simple method for bone marrow. Specimens are withdrawn from sternal punctures with a plastic syringe and left undisturbed in the syringe until clotted. The clot is removed gently and placed in a small vial containing the fixative.

### 3.6.2 FIXATION IN SUSPENSION IN MEDIUM

For many isolated cells good fixation is not obtained by adding the fixative to a pellet since the cells are affected by the discrepancy between the composition of the fixative and of their growth or suspending medium. Better fixation is often achieved by adding the fixative while the cells are still in suspension in medium.

The cell suspension is placed in a centrifuge tube and an equal volume (or more) of the fixative is added and mixed with the cells by gently shaking or tipping the tube. Double strength fixative can be used (e.g. Glauert and Thornley 1966) so that the final concentration of the fixative is the same as that used for pieces of tissue.

A few workers have used larger volumes of fixative than of cell suspension (e.g. Behnke 1968), while others have used small quantities of concentrated fixative (e.g. Weinbaum et al. 1970). Some typical examples of fixation of cells in suspension in medium are listed in Table 3.3 (see also Table 3.2).

The following fixation method was used successfully by Roth et al. (1970) to fix both the cell body and the axopodia of a helizoan. A glutaraldehyde fixative (12% glutaraldehyde in 0.033 M phosphate buffer, pH 5.3, 6.6 or 7.2, containing $2 \times 10^{-5}$ M magnesium sulphate and $2 \times 10^{-3}$ M sucrose) was added slowly to an equal volume of culture medium containing the organisms and allowed to remain for 30 sec. Then an amount of osmium tetroxide fixative equal to the glutaraldehyde fixative plus culture medium was added and the cells fixed for a further 20–30 min in the final mixed fixative, consisting of 3% glutaraldehyde and 0.5% osmium tetroxide in

TABLE 3.3

Fixation of isolated cells in suspension in medium

| Type of cell | Suspending medium | Ratio medium:fixative | Fixative | | pH | Temp. | Time | Typical reference |
|---|---|---|---|---|---|---|---|---|
| Platelets | Plasma | 1:4 | 4% G | 0.1 M cac | 6.5–6.8 | r.t. | 1 hr | Behnke (1968) |
| Mast cells | Peritoneal fluid | | P + G | 0.1 M cac | 7.4 | r.t. | | Padawer (1968) |
| Leucocytes | Incubation medium | 1:3 | 3% G | 0.1 M cac or 0.1 M ph | 7.4 | r.t. or 4°C | 2 hr overnight | Zucker-Franklin et al. (1971) |
| Bone marrow and blood | Plasma | 1:1 | 0.5–4% P | 0.1 M ph | 7.32–7.45 | 4° or 25°C | 2 hr to overnight | Carson et al. (1972) |
| Bacteria | Growth medium | 10:1 | 1% OsO$_4$ | VeAc + CaCl$_2$ | 6.0 | r.t. | 5 min | Ryter and Kellenberger (1958) |
| Bacteria | Growth medium | 1:1 | 5% G | 0.18 M cac | 7.2 | cold | 1 hr | Glauert and Thornley (1966) |
| Bacteria | Growth medium | 10:1 | 2% G | 0.1 M cac | 6.9 | r.t. | | Swanson et al. (1971) |
| Bacteria | Growth medium | | concentrated G; final conc. 6% G. | 0.1 M ph | 7.0 | 4°C | overnight | Weinbaum et al. (1970) |

*Key:* G, glutaraldehyde; P, paraformaldehyde; cac, cacodylate; ph, phosphate; VeAc, veronal-acetate.

0.016 M phosphate buffer, containing $10^{-5}$ M magnesium sulphate and $10^{-3}$ M sucrose.

Cells in suspension in medium can also be fixed by exposing them to the vapours of the fixative (Jenkins 1964) by placing two open vials, containing the cells in a drop of medium and the fixative respectively, together in a closed container, such as a Petri dish, for a few minutes.

After fixation the cell suspension is centrifuged and the pellet processed in the same way as after fixation in a pellet (§ 3.6.1).

### 3.6.3 ENCAPSULATING METHODS FOR PELLETS

Only certain favourable pellets are cohesive enough to remain intact after the initial fixation; others tend to disintegrate when the buffer is added for washing and material is easily lost. When this is likely to happen the pellet is 'encapsulated' by surrounding it with a soft gel of agar, fibrin or bovine serum albumin. The pellet can be encapsulated at any stage before dehydration has begun, but it is advisable for it to be done as soon as possible and thus avoiding having to centrifuge the cells to a pellet at each stage.

### *Agar method I*

(Ryter and Kellenberger 1958)

1. Prepare a 2% solution of agar by dissolving the agar (see Appendix) in boiling distilled water or in medium. Pour the solution into a test tube while it is still molten, and place the tube in a water bath at 45 °C. At this temperature the agar just remains liquid.
2. Also place the centrifuge tube containing the pellet of fixed cells in the water bath at 45 °C.
3. Transfer a *small* drop of agar (approx. 0.03 ml) to the centrifuge tube with a *warm* pipette, and shake the tube gently to suspend the cells in the agar.
4. Immediately tilt the tube so that the agar runs down the side of the tube and forms a drop on a cool glass microscope slide.
5. After the agar has set (one or two minutes) cut the solidified agar containing the cells into small cubes, about 1 mm$^3$, with a sharp single-edged razor blade.

These cubes are then processed in the same way as pieces of a cohesive pellet (§ 3.6.1).

It is important to ensure that the agar remains warm until the moment it is poured onto the slide or it will start to set before it is thoroughly mixed with the cells. The amount of agar added should be kept as small as possible to avoid diluting the cells more than is necessary; if too much agar is added it will be found that a thin section through the agar contains only a very few cells.

### Agar method II

(Hirsch and Fedorko 1968)

1. Place the centrifuge tube containing the pellet of fixed and washed cells in a water bath at 50 °C.
2. Add a few drops of warm (50 °C) 2% agar to the pellet, using a warm pipette, and shake the tube to suspend the cells in the agar.
3. Centrifuge the tube containing the cells in agar at 750 g for 2 min in a carrier half-filled with hot tap water. The cells form a firm pellet at the bottom of the tube.
4. Cool the centrifuge tube in crushed ice to solidify the agar.
5. Add some 70% ethanol and leave the tube in ice for one hour or more.
6. Displace the pellet by carefully pipetting; remove it from the centrifuge tube and cut it into small cubes with a sharp, single-edged razor blade.

It will be noted that Hirsch and Fedorko (1968) recommend handling the agar at a higher temperature (50 °C) than that used by Ryter and Kellenberger (1958) (45 °C). There is less chance that the agar will set too soon at the higher temperature, but it is possible that the structure of the specimen will be affected, even after fixation, if the temperature rises too high. In general, it is advisable to keep the agar as cool as possible without allowing it to solidify.

Winters and Slade (1971) described a method of staining agar with Schiff's solution (leuco basic fuchsin) to make the small cubes more easily visible.

### Fibrin clot method

Charret and Fauré-Fremiet (1967) showed that it is possible to use a fibrin clot instead of agar for encapsulation with the advantages that it is not necessary to raise the temperature and that the clot is more readily penetrated by dehydrating and embedding solutions than agar. The method is based on the ability of thrombin to transform a solution of fibrinogen into a fibrin clot at room temperature. Fibrin is compatible with most buffer systems, except those containing cacodylate (Sicko and Arnold 1971). By adjusting the fibrinogen content clots can be prepared with densities ranging from an open, mucus-like network (200 mg per 100 ml) to a hard, agar-like gel (1.5%). Sicko and Arnold (1971) recommend the use of freshly prepared 0.7% bovine fibrinogen for animal cell suspensions and blue green algae. Both thrombin and fibrinogen are also available commercially (see Appendix).

Furtado (1970) studied the conditions for the application of the method and developed the following technique for encapsulating pellets:

1. Dilute the commercial thrombin solution with sterile distilled water and store at a concentration of 1000 units/ml in 1 ml aliquots in vials. Dilute to a concentration of 50–100 units/ml before use.
2. Prepare a solution containing 300–350 mg fibrinogen, 160 mg sodium citrate and 850 mg sodium chloride in 100 ml distilled water.
3. Remove the washing solution from a pellet of fixed and washed cells in a centrifuge tube and add 0.2–0.3 ml of fibrinogen solution. Stir and centrifuge again to concentrate the cells, if necessary.
4. Add an equal volume of thrombin solution slowly to the supernatant fluid with a pipette. Twist a sharpened applicator stick in the fibrinogen solution until a veil-like clot becomes evident. Then twist the stick in the pellet, until the cell mass is entrapped at the tip of the stick.
5. Use the stick to transfer the clot through the subsequent steps in the preparative procedure. The clot is cut into small pieces when in 70% ethanol.

### Bovine serum albumin method

Bovine serum albumin can also be used to encapsulate pellets without raising the temperature. The albumin is gelled by the addition of glutaraldehyde, and Shands (1968) recommends the following method:

1. Prepare a 2% solution of bovine serum albumin in 0.05 M Tris buffer, pH 7.5, and filter the solution through a 0.45 $\mu$m pore-size Millipore filter.
2. Resuspend the pellet of fixed cells (or other particles) in 0.25 ml of 2% bovine serum albumin in the same Tris buffer and transfer the suspension to a small cellulose centrifuge tube.
3. Add one drop of 25% (distilled) glutaraldehyde and mix it with the suspension. Centrifuge the suspension immediately in a swinging bucket type rotor. A gel forms in about 5 min and is easily detectable as the bovine serum albumin becomes opaque.
4. Remove the stopper from the tube, and cut the tube and its contents into slices about 1 mm thick. Place the slices on filter paper to remove excess moisture.
5. Cut the surrounding rims of cellulose to release the discs of gel. Dehydrate and embed these discs by standard methods.

Shands (1968) used this method successfully with macrophages and bacteria, as well as smaller particles, such as mitochondria and antigen-antibody precipitates. The only disadvantage of the method is that block staining of the pieces of gel may produce a background granularity.

### 3.6.4  ENCAPSULATING METHODS FOR ISOLATED CELLS

Isolated cells, such as protozoa, that are large enough to be handled individually can also be encapsulated to simplify the subsequent dehydration and embedding.

*Agar method*

In the technique developed by De Haller et al. (1961) a fixed cell is transferred with a pipette to a drop of 2% liquid agar on a glass slide on a hot plate at 45 °C. The authors recommend that the agar be dissolved in the buffer used for the fixative. The cell is viewed under a microscope and orientated as required, while the agar is still liquid, using a fine glass rod. The glass slide is then placed in a humid atmosphere in a Petri dish and the agar is allowed to set. A small block of agar, containing the cell, is then cut out with a small scapel or sharp razor blade, and this block is then dehydrated and embedded. The cells are difficult to see in some embedding media, such as Vestopal, and De Haller et al. (1961) suggest placing a small pointer made from aluminium in the agar pointing at the cell.

*Albumen method*

Isolated cells can also be encapsulated in a fibrin clot using the same method as for pellets (§ 3.6.3), or in albumen, the more fluid part of natural egg white. Sawicki and Lipetz (1971) proposed the following method:
1. Prepare gelatinized glass slides by immersing them in a 0.5% aqueous solution of gelatin for 3 min and then allowing them to dry.
2. Place small drops (approx. 0.02 ml) of albumen on the slides.
3. Transfer fixed cells to the microdroplets of albumen using a micropipette. One or more cells may be introduced into each drop.
4. Invert the slide for about 1 min. The hanging drops elongate and the cells move to the bottom.
5. Expose the drops (still inverted) to the vapours of absolute ethanol or 40% formaldehyde in a closed dish. The drops coagulate in about 5 min and become opaque.
6. Preparations coagulated by formaldehyde are washed in water and then placed in 70% ethanol; the others are transferred directly to 70% ethanol. Stand the slides vertically in the ethanol in a Coplin jar.

7. Dehydrate the drops in increasing concentrations of ethanol, and remove the drops from the slides and place them in small vials before impregnation with the embedding medium. The hardened drops are easily removed from the gelatinized slide.

3.6.5 COLLECTING CELLS ON MILLIPORE FILTERS

A convenient method of collecting isolated cells for fixation is to use a Millipore membrane filter. Filters with various pore sizes are available (see Appendix), 0.45 μm being suitable for cell cultures and some bacteria, 0.22 μm for smaller bacteria and bacterial spheroplasts and 50 nm for viruses. Cells are collected either by filtering the suspension through the Millipore membrane in the usual way or a sterile filter is placed in a Petri dish and the concentrated suspension placed gently in the centre. The filter is then incubated at 37 °C for 2–3 hr to allow the cells to attach firmly (McCombs et al. 1968). The cells on the filter are fixed and dehydrated in the same way as monolayers of cells grown on filters or other substrata (§ 3.5), the filter being cut into pieces before impregnation with embedding medium. Propylene oxide may dissolve the filter and so xylene or toluene is used instead (see § 5.8.2d).

## 3.7 Cell fractions

Subcellular fractions, such as microsomes, mitochondria or bacterial cell walls, are handled in the same way as isolated cells, but higher speeds of centrifugation are usually required because of the smaller size of the particles. Adequate fixation is often obtained by centrifuging the fraction to a pellet and then adding the fixative as described in § 3.6.1. Plastic centrifuge tubes should be used for easy removal of the fixed pellet. Some membrane-bounded particles, such as mitochondria which are sensitive to the osmotic conditions, show signs of damage when fixed in this way, and it is then advisable to fix the particles while they are still in suspension in the isolation medium. The fraction is fixed by mixing it with an equal volume (or more) of a fixative, which contains the same concentration of sucrose as the suspending medium, as in the method for suspensions of cells (§ 3.6.2). Alternatively, sufficient purified, concentrated glutaraldehyde is added to the suspension to give a final concentration of glutaraldehyde of about 1%.

Millonig and Marinozzi (1968) have shown that glutaraldehyde is preferable to osmium tetroxide for the fixation of fractions suspended in sucrose because sucrose inhibits proper fixation of proteins by osmium tetroxide. They also recommend first fixing pellets from cell fractions in glutaraldehyde vapour for several minutes before adding the liquid fixative. The composition of the glutaraldehyde fixative does not appear to be critical, concentrations of glutaraldehyde ranging from 0.5–3.0% in 0.1 M cacodylate buffer being customary, calcium chloride (2–3 mM) being added for the fixation of membraneous particles. Fractions in the form of a pellet are fixed for 30 min to 1 hr, the time depending on the size of the pellet, while much shorter times (e.g. 10 min) are required for fixation in suspension. These fractions are centrifuged to a pellet after fixation, before washing and post-fixation with osmium tetroxide. If the pellet tends to disintegrate in the washing solution, it is encapsulated by one of the methods described for pellets of isolated cells (§ 3.6.3).

The pellet is removed from the centrifuge tube and cut into small cubes or strips for post-fixation in osmium tetroxide, care being taken that samples are taken from all levels in the pellet so that the heterogeneity of the sample is fully explored. Whenever possible, it is advisable to cut the pellet into strips aligned in the direction of the axis of the centrifuge tube and then to section these strips longitudinally. In this way the different layers of the pellet are visible in the final section.

Baudhuin et al. (1967) found that the heterogeneity in some pellets does not occur solely in the direction of the centrifugal force, but also extends along the surface of the preparation. As a result particles of different types occur with differing frequencies in the centre and in the periphery of the preparation. To overcome this problem they developed a method of preparing very thin (about 10 μm) pellicles of packed particles in which heterogeneity is solely in a direction perpendicular to the surface of the pellicle. The method involves filtration on a Millipore filter under pressure and is somewhat complicated. It has the great advantage that sections covering the whole depth of the pellicle can be photographed in a single field in the electron microscope.

### 3.8    Methods for very small quantities of material

When only small quantities of cells or cell fractions are available it is necessary to use correspondingly small centrifuge tubes in order to obtain a compact pellet that is not spread in a thin layer over the bottom of the tube. Malamed (1963) suggested using disposable, polyethylene tubes of 0.4 ml capacity such as those that fit into the high speed Microfuge (Beckman Instruments, Spinco Division, see Appendix). A small pellet is transferred on the tip of a

Pasteur pipette to a microcentrifuge tube about two-thirds filled with fixative. After the required fixation time, a pellet is formed by centrifugation (Fig. 3.3) and the fixative is withdrawn. The pellet is released by cutting the tube transversely just above the top of the pellet with a sharp razor blade, making a similar cut just beneath the pellet and then dislodging the pellet by poking a wooden toothpick into the lower, smaller hole. The pellet is then handled in the same way as larger pellets (§ 3.6.1).

Tissue particulate suspended in fluid    Centrifuge 5 minutes at 10,000 x g    Withdraw fixative    Cut Microfuge tube with razor blade    Transfer pellet to vial with blunt needle

Fig. 3.3. Schematic representation of the microcentrifugal method for processing material for electron microscopy (from Malamed 1963).

Another method of handling very small quantities of material is described by Marikovsky and Danon (1963, 1966). The fraction is placed in a poly-ethylene capillary tube (1 mm internal diameter, 70 mm long) containing a column of phthalate esters. The capillary tube is sealed at both ends and then centrifuged in a microhematocrit centrifuge at 12,000 $g$ for 15 min. The fraction separates into different layers and the required layers are obtained by cutting the tube into segments with a razor blade. These layers are fixed, washed and dehydrated, the fraction being expelled from the capillary tube in 70% ethanol (or acetone) using a thin wire.

A third method of handling very small amounts of material is to carry out the whole preparative procedure, from fixation to embedding, in a pyramidal-tipped plastic capsule (BEEM, see Appendix) (§ 5.2). The material is centrifuged to a pellet for each change of solution and care has to be taken to resuspend the sample properly, particularly during infiltration with the embedding medium which should be chosen to have low viscosity (e.g. Spurr epoxy resin) (§ 5.3.2a). Demaree (1970) suggests tying a thread to the capsule so that it can be easily lowered into a tapered centrifuge tube, the thread being taped to the outside of the tube. Alternatively, a special carrier for holding a number of capsules can be used. Kondo and Takemura (1965) describe a holder made out of cork, while Blackburn (1968) recommends a carrier made from brass. Centrifuge tubes modified to hold BEEM capsules are available commercially (see Appendix).

## 3.9   Special fixation schedules and methods

The majority of specimens are well fixed by the methods already described with a standard fixation schedule, similar to those outlined in § 3.2. Certain specimens or chemical components require the application of modified fixation methods, and some of these are outlined briefly in this section.

### 3.9.1   LIPIDS

Lipids are poorly preserved by routine fixation methods (Stein and Stein 1971) so that a high proportion of the lipid may be lost during the subsequent dehydration (Chapter 4). Alterations in the dehydration procedure are the best way to preserve lipids, but various modifications to the standard fixation procedure of glutaraldehyde followed by osmium tetroxide help to reduce the amount of lipid lost, including:

i.    Addition of calcium chloride to the glutaraldehyde, and possibly also the subsequent osmium tetroxide, fixative to reduce the extraction of *phospholipids* (e.g. Mitchell 1969).

ii.   Addition of digitonin to the primary aldehyde, or aldehyde plus osmium tetroxide, fixative to reduce extraction of *free cholesterol* (§ 2.8.5).

iii.  Fixation by Baker's (1946) formal-calcium procedure, as modified by Casley-Smith (1963). Specimens are fixed firstly with 4% formaldehyde plus 1% calcium chloride, in 0.03 M veronal–acetate buffer, pH 7.4, containing 7.5% sucrose, for 18 hr at 3 °C, and then secondly in 5% potassium dichromate plus 1% calcium chloride for 18 hr at 20 °C and then for 24 hr at 60 °C. This fixation method improves the preservation of *butter* and *maize oil* (Casley-Smith 1963). The preservation of *phospholipids* is no better than after routine fixation in glutaraldehyde followed by osmium tetroxide (Cope and Williams 1969).

iv.   Tricomplex fixation (see § 2.8.3) for the preservation of *labile lipids*.

### 3.9.2   CELLS WITH THICK WALLS

Thick-walled cells, such as yeasts, and certain other specimens, such as sea urchin eggs, are difficult to fix well because the fixative does not penetrate easily. Various techniques have been suggested for aiding the penetration of the fixative, including:

i. Use of a fixative containing acrolein, the most rapidly penetrating of the aldehyde fixatives (§ 2.6.1), or potassium permanganate (§ 2.7.1), (e.g. Wallace et al. 1968).

ii. Mechanical breakage of the cell before or during fixation by cutting (Bluemink 1970) or by shaking with glass beads. For example, Damsky et al. (1969) poured a yeast culture over ice into flasks containing sufficient paraformaldehyde or glutaraldehyde to give a final concentration of fixative of 1 or 2.5% respectively, and after 20 min or less ruptured the cells in a Braun glass bead mill. Small pieces of the cell wall were removed but the cell outlines were not destroyed. The broken cells were fixed for a further period in the aldehyde fixative and then with osmium tetroxide.

### 3.9.3 BACTERIA

It is difficult to obtain good fixation of bacteria by primary fixation with osmium tetroxide because of the labile nature of the nuclear material, which is associated with relatively little protein and which is not surrounded by a nuclear membrane. In consequence Ryter and Kellenberger (1958) developed a special method for the fixation of bacteria (in the period before the introduction of aldehyde fixatives), and this schedule is still widely used. The method was designed with the specific aim of stabilising the nuclear material.

#### *Ryter-Kellenberger standard fixation for bacteria*

1. *Pre-fixation.* Mix 30 ml of bacteria suspended in tryptone culture medium (1% bacto-tryptone, Difco, and 0.5% sodium chloride) in a centrifuge tube with 3 ml of Ryter-Kellenberger veronal-acetate buffered osmium tetroxide fixative (§ 2.3.5), and centrifuge immediately for 5 min at 1800 g.

2. *Main fixation.* Resuspend the pellet in 1.0 ml of Ryter-Kellenberger fixative and 0.1 ml of tryptone medium, and leave overnight (about 16 hr) at room temperature.

3. Dilute the suspension with 8 ml of Ryter-Kellenberger veronal-acetate buffer (§ 2.2.7) and centrifuge for 5 min at 1800 g.

4. Resuspend the pellet in about 0.03 ml of warm agar and put the drop on a microscope slide (§ 3.6.3). Cut the hardened clot into small cubes, approx. 1 mm³.

5.   Place the cubes in 0.5% uranyl acetate in Ryter-Kellenberger veronal-acetate buffer for 2 hr at room temperature. Dehydrate and embed.

This fixation schedule preserves the nuclear material of bacteria in a finely fibrillar state, and also results in a clear demonstration of membranes, due to the presence of calcium and uranyl ions which help to preserve lipids (Silva et al. 1968; Burdett and Rogers 1970). Recent studies suggest that 'pre-fixation' in the presence of a low concentration of osmium tetroxide may lead to a reorganisation of the membranes of mesosomes (Silva 1971). In general, this step is omitted, the pellet of bacteria, or cells on Millipore filters or agar, being placed directly in full-strength fixative.

Improved results are also obtained by fixing the bacteria first in a glutaraldehyde fixative, either in suspension or in a pellet, using a similar fixation schedule as for mammalian cells. This primary glutaraldehyde fixation is followed by fixation for 1-16 hr in Ryter-Kellenberger osmium tetroxide fixative and then by fixation in uranyl acetate before dehydration and embedding.

### 3.9.4   MARINE ORGANISMS

Sea water is added to fixatives for marine organisms. Osmium tetroxide is dissolved directly in sea water (e.g. Afzelius 1956), while glutaraldehyde fixatives are prepared by mixing double-strength glutaraldehyde (e.g. 6%) with an equal volume of sea water, or by buffering the fixative with sodium cacodylate dissolved in filtered sea water (e.g. Anderson and Personne 1970).

Potassium permanganate fixatives have also been used for the fixation of marine organisms. Doggenweiler and Heuser (1967) recommend a 1.5% solution of potassium permanganate in artificial sea water for the fixation (for 3 hr at 4 °C) of the prawn nerve sheath, while Martin and Rosenberg (1968) used 0.6% potassium permanganate in Tris-buffered sea water at pH 7.5 to 8.0 for the fixation (for $\frac{1}{2}$ to 2 hr) of squid giant axon fibres.

### 3.9.5   ORGANISMS LIVING IN EXTREME CONDITIONS

Uni- or multi-cellular organisms which live in environments of very low or very high ionic strength, or at alkaline or acid pH, are very difficult to fix. It is paramount that the fixative has a pH near neutrality and that the concentration of electrolytes matches that of the interior of the cell (Millonig 1973).

### 3.9.6   SPECIAL FIXATION SCHEDULES

The standard fixation schedules outlined in § 3.2 are suitable for the great majority of specimens. Some modified schedules for special purposes have been described in this chapter, and these are listed in Table 3.4, together with a selection of fixation schedules which are not widely used but may prove to be of value in special applications.

TABLE 3.4

Fixation schedules

| Sequence of fixatives | Type of specimen | Typical reference |
|---|---|---|
| *Standard fixation schedules* | | |
| 1. Aldehydes + $Ca^{2+}$; $OsO_4$ | Various | Karnovsky (1965) |
| 2. Aldehydes; $OsO_4$; UA | Various | Farquhar and Palade (1965) |
| *Fixation schedules for special purposes* | | |
| 3. F + $Ca^{2+}$; $K_2Cr_2O_7$ + $Ca^{2+}$ | Lipids | Casley-Smith (1963) |
| 4. G + $OsO_4$; UA | Isolated cells and monolayers | Hirsch and Fedorko (1968) |
| 5. $OsO_4$ + $K_2Cr_2O_7$ | Various | Dalton (1955) |
| 6. $OsO_4$; UA | Bacteria | Ryter and Kellenberger (1958) |
| 7. $KMnO_4$ | Plants | Mollenhauer (1959) |
| 8a. $OsO_4$ in sea water | Marine | Afzelius (1956) |
| 8b. G in sea water | organisms | Anderson and Personne (1970) |
| 8c. $KMnO_4$ in sea water | Marine organisms | Doggenweiler and Heuser (1967) |
| *Fixation schedules not widely used* | | |
| 9. G or F; $K_2Cr_2O_7$; $OsO_4$ | Isolated cells | Sugihara et al. (1966) |
| 10. G; $Pb(NO_3)_2$ + $K_3Fe(CN)_6$; $OsO_4$ | Lung lipids | Dermer (1969) |
| 11. G; $KMnO_4$ | Plants | Scala et al. (1968) |
| 12. G + A; $KMnO_4$ | Plants | Hayat (1970) |
| 13. Aldehyde; $OsO_4$ + $K_3Fe(CN)_6$ | Various | Karnovsky (1971) |
| 14. G + $OsO_4$; $OsO_4$ | Various | Franke et al. (1969) |
| 15. G + $OsO_4$; G | Amphibian | Bluemink (1970) |
| 16. G + F + trinitro compounds | Steroid-secreting tissues | Ito and Karnovsky (1968) |
| 17. $OsO_4$; F | Connective tissues | Ross and Klebanoff (1969) |
| 18. $OsO_4$; G; UA | Brain | Kanaseki and Kadota (1969) |
| 19. $OsO_4$ + $K_2Cr_2O_7$; UA | Tissue culture | Zeigel et al. (1969) |
| 20. $KMnO_4$; $K_2Cr_2O_7$ + UN | Yeast | Wallace et al. (1968) |
| 21. $KMnO_4$ + $Ca^{2+}$; $K_2Cr_2O_7$ | Yeast | Kawakami (1961) |

*Key*: UA, uranyl acetate; F, formaldehyde; G, glutaraldehyde; A, acrolein; UN, uranyl nitrate.

REFERENCES

Afzelius, B. A. (1956), The ultrastructure of the cortical granules and their products in the sea urchin egg as studied with the electron microscope, Expl Cell Res. *10*, 257.

Anderson, D. R. (1965), A method of preparing peripheral leucocytes for electron microscopy, J. Ultrastruct. Res. *13*, 263.

Anderson, R. and J. Cronshaw (1970), Sieve element pores in *Nicotiana* pith culture, J. Ultrastruct. Res. *32*, 458.

Anderson, W. A. and P. Personne (1970), The localization of glycogen in the spermatozoa of various invertebrate and vertebrate species, J. Cell Biol. *44*, 29.

Atkinson, A. W., B. E. S. Gunning and P. C. L. John (1972), Sporopollenin in the cell wall of *Chlorella* and other algae: ultrastructure, chemistry and incorporation of [$^{14}$C] acetate, studied in synchronous culture, Planta *107*, 1.

Bain, J. M. and D. W. Gove (1971), Rapid preparation of plant tissues for electron microscopy, J. Microscopy *93*, 159.

Baker, J. R. (1946), The histochemical recognition of lipine, Q. Jl Microsc. Sci. *87*, 441.

Baldwin, K. M. (1970), The fine structure and electrophysiology of heart muscle cell injury, J. Cell Biol. *46*, 455.

Baudhuin, P., P. Evrard and J. Berthet (1967), Electron microscopic examination of subcellular fractions. 1. The preparation of representative samples from suspensions of particles, J. Cell Biol. *32*, 181.

Behnke, O. (1968), Electron microscopical observations on the surface coating of human blood platelets, J. Ultrastruct. Res. *24*, 51.

Bisalputra, T. and A. A. Bisalputra (1969), The ultrastructure of chloroplast of a brown alga *Sphacelaria* sp. 1. Plastid DNA configuration – the chloroplast genophore, J. Ultrastruct. Res. *29*, 151.

Blackburn, W. R. (1968), Carrier centrifuge tube for gelatin and polyethylene capsules, J. Ultrastruct. Res. *23*, 362.

Bluemink, J. G. (1970), The first cleavage of the amphibian egg, J. Ultrastruct. Res. *32*, 142.

Bodian, D. (1970), An electron microscopic characterization of classes of synaptic vesicles by means of controlled aldehyde fixation, J. Cell. Biol. *44*, 115.

Bohmann, S. O. and A. B. Maunsbach (1970), Effects on tissue fine structure of variations in colloid osmotic pressure of glutaraldehyde fixatives, J. Ultrastruct. Res. *30*, 195.

Brightman, M. W. and T. S. Reese (1969), Junctions between intimately apposed cell membranes in the vertebrate brain, J. Cell Biol. *40*, 648.

Burdett, I. D. J. and H. J. Rogers (1970), Modification of the appearance of mesosomes in sections of *Bacillus licheniformis* according to the fixation procedures, J. Ultrastruct. Res. *30*, 354.

Carson, F., J. A. Lynn and J. H. Martin (1972), Ultrastructural effect of various buffers, osmolality, and temperature on paraformaldehyde fixation of the formed elements of blood and bone marrow, Tex. Rep. Biol. Med. *30*, 125.

Casley-Smith, J. R. (1963), Some observations on the fixation and staining of lipids, Jl R. microsc. Soc. *81*, 235.

Charret, R. and E. Fauré-Fremiet (1967), Technique de rassemblement de microorganismes: préinclusion dans un caillot de fibrine, J. Microscopie *6*, 1063.

Cope, G. H. and M. A. Williams (1969), Quantitative studies on the preservation of choline and ethanolamine phosphatides during tissue preparation for electron microscopy. II. Other preparative methods, J. Microscopy *90*, 47.

Dalton, A. J. (1955), A chrome–osmium fixative for electron microscopy, Anat. Rec. *121*, 281.

Damsky, C. H., W. M. Nelson and A. Claude (1969), Mitochondria in anaerobically-grown lipid-limited brewer's yeast, J. Cell Biol. *43*, 174.

De Haller, G., C. F. Ehret and R. Naef (1961), Technique d'inclusion et d'ultramicrotomie, destinée à l'étude du développement des organelles dans une cellule isolée, Experientia *17*, 524.

Demaree, R. S. (1970), Isolation of avian trypanosomes by selective centrifugation and subsequent processing for electron microscopy, Stain technol. *45*, 304.

Dermer, G. B. (1969), The fixation of pulmonary surfactant for electron microscopy. 1. The alveolar surface lining layer, J. Ultrastruct. Res. *27*, 88.

Doggenweiler, C. F. and J. E. Heuser (1967), Ultrastructure of the prawn nerve sheath. Role of fixative and osmotic pressure in vesiculation of thin cytoplasmic laminae, J. Cell Biol. *34*, 407.

Falk, H. (1969), Rough thylakoids: polysomes attached to chloroplast membranes, J. Cell Biol. *42*, 582.

Farquhar, M. G., D. F. Bainton, M. Baggiolini and C. De Duve (1972), Cytochemical localization of acid phosphatase activity in granule fractions from rabbit polymorphonuclear leukocytes, J. Cell Biol. *54*, 141.

Farquhar, M. G. and G. E. Palade (1965), Cell junctions in amphibian skin, J. Cell Biol. *26*, 263.

Franke, W. W., S. Krein and R. M. Brown (1969), Simultaneous glutaraldehyde-osmium tetroxide fixation with postosmication. An improved fixation for electron microscopy of plant and animal cells, Histochemie *19*, 162.

Furtado, J. S. (1970), The fibrin clot: a medium for supporting loose cells and delicate structures during processing for microscopy, Stain technol. *45*, 19.

Gardner, H. A., G. Simon and M. D. Silver (1969), The fine structure of rabbit platelets fixed in acetaldehyde, Anat. Rec. *163*, 509.

Gil, J. and E. R. Weibel (1969–70), Improvements in demonstration of lining layer of lung alveoli by electron microscopy, Respiration Physiol. *8*, 13.

Glauert, A. M. and D. A. Hopwood (1960), The fine structure of *Streptomyces coelicolor*. 1. The cytoplasmic membrane system, J. biophys. biochem. Cytol. *7*, 479.

Glauert, A. M. and M. J. Thornley (1966), Glutaraldehyde fixation of Gram-negative bacteria, Jl R. microsc. Soc. *85*, 449.

Hayat, M. E. (1970), Principles and techniques of electron microscopy, Vol. 1, Biological applications (Van Nostrand Reinhold, New York).

Hirsch, J. G. and M. E. Fedorko (1968), Ultrastructure of human leukocytes after simultaneous fixation with glutaraldehyde and osmium tetroxide and 'postfixation' in uranyl acetate, J. Cell Biol. *38*, 615.

Hoch, H. C. and J. E. Mitchell (1972), The ultrastructure of zoospores of *Aphanomyces euteiches* and of their encystment and subsequent germination, Protoplasma *75*, 113.

Ito, S. (1962), Post-mortem changes of the plasma membrane, Proc. 5th Int. Congr. Electron Microscopy, Philadelphia *2*, L-5.

Ito, S. and M. J. Karnovsky (1968), Formaldehyde-glutaraldehyde fixatives containing trinitro compounds, J. Cell Biol. *39*, 168a.

Ito, S. and R. J. Winchester (1963), The fine structure of the gastric mucosa in the bat, J. Cell Biol. *16*, 541.

Jenkins, R. A. (1964), Osmium tetroxide vapor fixation of protozoan cells, J. Cell Biol. *23*, 46A.

Kanaseki, T. and K. Kadota (1969), The 'vesicle in a basket', J. Cell Biol. *42*, 202.

Karnovsky, M. J. (1965), A formaldehyde–glutaraldehyde fixative of high osmolality for use in electron microscopy, J. Cell Biol. *27*, 137A.

Karnovsky, M. J. (1967), The ultrastructural basis of capillary permeability studied with peroxidase as a tracer, J. Cell Biol. *35*, 213.

Karnovsky, M. J. (1971), Use of ferrocyanide-reduced osmium tetroxide in electron microscopy, Proc. 14th Ann. Meeting Amer. Soc. Cell Biol., p. 146.

Kawakami, N. (1961), A permanganate–chrome fixative and lead acetate staining for electron microscopy of microorganisms, J. Electron Microscopy 10, 14.

Kondo, K. and K. Takemura (1965), Application of epoxy resin embedding method for free cells, J. Electron Microscopy 14, 50.

Lewis, P. R., D. P. Knight and M. A. Williams (1974), Staining methods for thin sections, in: Practical methods in electron microscopy, A. M. Glauert, ed. (North-Holland, Amsterdam).

López-Sáez, J. F., G. Giménez-Martin and M. C. Risueño (1966), Fine structure of the plasmodesm, Protoplasma 61, 81.

Lumb, W. V. (1963), Small animal anesthesia (Lea and Febiger, Philadelphia).

Malamed, S. (1963), Use of a microcentrifuge for preparation of isolated mitochondria and cell suspensions for electron microscopy, J. Cell Biol. 18, 696.

Marikovsky, Y. and D. Danon (1963), A micromethod for fixation and embedding of separated blood cell fractions for electron microscopy, J. Ultrastruct. Res. 18, 176.

Marikovsky, Y. and D. Danon (1966), A micromethod for fixation and embedding of blood cells separated according to their density, Proc. 6th Int. Congr. Electron Microscopy, Kyoto 2, 31.

Marks, I. and L. G. Briarty (1970), A syringe-filter for processing specimens through fluids used in plastic embedding, Stain technol. 45, 36.

Martin, R. and P. Rosenberg (1968), Fine structural alterations associated with venom action on squid giant axon fibers, J. Cell Biol. 36, 341.

Maunsbach, A. B. (1966), Absorption of $I^{125}$-labeled homologous albumin of rat kidney proximal tubule cells, J. Ultrastruct. Res. 15, 197.

Maunsbach, A. B., S. C. Madden and H. Latta (1962), Variations in fine structure of renal tubular epithelium under different conditions of fixation, J. Ultrastruct. Res. 6, 511.

McCombs, R. M., M. Benyesh-Melnick and J. P. Brunschwig (1968), The use of Millipore filters in ultrastructural studies of cell cultures and viruses, J. Cell Biol. 36, 231.

Millonig, G. (1973), Personal communication, Laboratory of Molecular Embryology, University of Naples.

Millonig, G. and V. Marinozzi (1968), Fixation and embedding in electron microscopy, in: Adv. Optical and Electron Microscopy, Vol. 2, R. Barer and V. E. Cosslett, eds. (Academic Press, New York and London), p. 251.

Mitchell, C. D. (1969), Preservation of the lipids of the human erythrocyte stroma during fixation and dehydration for electron microscopy, J. Cell Biol. 40, 869.

Mollenhauer, H. H. (1959), Permanganate fixation of plant cells, J. biophys. biochem. Cytol. 6, 431.

Mollenhauer, H. H. (1967). The fine structure of mucilage secreting cells of Hibiscus esculentus pods, Protoplasma 63, 353.

Nobel, P. S., S. Murakami and A. Takamiya (1966), Localization of light-induced strontium accumulation in spinach chloroplasts, Plant and Cell Physiol. 7, 263.

Öpik, H. (1972), Some observations on coleoptile cell ultrastructure in ungerminated grains of rice (Oryza sativa L), Planta 102, 61.

Padawer, J. (1968), Uptake of colloidal thorium dioxide by mast cells, J. Cell Biol. 40, 747.

Palade, G. E. (1956), The fixation of tissues for electron microscopy, Proc. 3rd Int. Congr. Electron Microscopy, London, p. 129.

Palfrey, A. J. and D. V. Davies (1966), The fine structure of chondrocytes, J. Anat. 100, 213.

Pease, D. C. (1964), Histological techniques for electron microscopy, 2nd edition (Academic Press, New York and London).

Peters, A., C. C. Proskauer and I. R. Kaiserman-Abramof (1968), The small pyramidal neuron of the rat cerebral cortex, J. Cell Biol. 39, 604.

Pickett-Heaps, J. D. and D. H. Northcote (1966), Organization of microtubules and endoplasmic reticulum during mitosis and cytokenesis in wheat meristems, J. Cell Sci. *1*, 109.

Porter, K. R. and F. Kallman (1953), The properties and effects of osmium tetroxide as a tissue fixative with special reference to its use for electron microscopy, Expl Cell Res. *4*, 127.

Reese, T. S. and M. J. Karnovsky (1967), Fine structural localization of a blood-brain barrier to exogenous peroxidase, J. Cell Biol. *34*, 207.

Rhodin, J. (1954), Correlation of ultrastructural organization and function in normal and experimentally changed proximal convoluted tubule cells of the mouse kidney, Thesis, Stockholm.

Robards, A. W. (1968), On the ultrastructure of differentiating secondary xylem in willow, Protoplasm *65*, 449.

Robertson, J. D. (1963), The occurrence of a subunit pattern in the unit membranes of club endings in Maunther cell synapses in goldfish brain, J. Cell Biol. *19*, 201.

Ross, R. (1972), Personal communication, Department of Pathology, University of Washington, Seattle.

Ross, R. and S. J. Klebanoff (1967), Fine structural changes in uterine smooth muscle and fibroblasts in response to estrogen, J. Cell Biol. *32*, 155.

Roth, L. E., D. J. Pihlaja and Y. Shigenaka (1970), Microtubules in the helizoan axopodium, J. Ultrastruct. Res. *30*, 7.

Ryter, A. and E. Kellenberger (1958), Etude au microscope électronique de plasmas contenant de l'acide désoxyribonucléique, Z. Naturf. *13*, 597.

Sawicki, W. and J. Lipetz (1971), Albumen embedding and individual mounting of one or many mammalian ova on slides for fluid processing, Stain technol. *46*, 261.

Scala, J., D. Schwab and E. Simmons (1968), The fine structure of the digestive gland of Venus's flytrap, Am. J. Bot. *55*, 649.

Schulz, R. and W. Jensen (1968), *Capsella* embryogenesis: the early embryo, J. Ultrastruct. Res. *22*, 376.

Schultz, R. L. and N. M. Case (1970), A modified aldehyde perfusion technique for preventing certain artifacts in electron microscopy of the central nervous system, J. Microscopy *92*, 69.

Shands, J. W. (1968), Embedding free-floating cells and microscopic particles: serum albumin coagulum – epoxy resin, Stain technol. *43*, 15.

Sicko, L. M. and W. J. Arnold (1971), Fibrin clots for electron microscopy and cytochemistry of minute biological samples, Proc. 29th Ann. Meeting EMSA, p. 472.

Silva, M. T. (1971), Changes induced in the ultrastructure of the cytoplasmic and intracytoplasmic membranes of several Gram-positive bacteria by variations in $OsO_4$ fixation, J. Microscopy *93*, 227.

Silva, M. T., F. C. Guerra and M. M. Magalhães (1968), The fixative action of uranyl acetate in electron microscopy, Experientia *24*, 1074.

Sjöstrand, F. S. and L.-G. Elfvin (1962), The layered asymmetric structure of the plasma membrane in the exocrine pancreas cells of the cat, J. Ultrastruct. Res. *7*, 504.

Smith, R. E. and M. G. Farquhar (1966), Lysosome function in the regulation of the secretory process in cells of the anterior pituitary gland, J. Cell Biol. *31*, 319.

Srivastava, L. M. and T. P. O'Brien (1966), On the ultrastructure of cambium and its vascular derivatives, I. Cambium of *Pinus strobus* L, Protoplasma *61*, 257.

Stein, O. and Y. Stein (1971), Light and electron microscopic radioautography of lipids: techniques and biological applications, Adv. Lipid Res. *9*, 1.

Sugihara, R., K. J. Lee, T. Sugioka and G. Yasuzumi (1966), Experiments on triple fixation for electron microscopy, Proc. 6th Int. Congr. Electron Microscopy, Kyoto *2*, 25.

Swanson, J., S. J. Kraus and E. C. Gotschlich (1971), Studies on gonococcus infections, J. exp. Med. *134*, 886.

Tormey, J. McD. (1963), Fine structure of the ciliary epithelium of the rabbit, with particular reference to 'infolded membranes', 'vesicles', and the effects of Diamox, J. Cell Biol. *17*, 641.

Trump, B. F., P. J. Goldblatt and R. E. Stowell (1962), An electron microscopic study of early cytoplasmic alteration in hepatic parenchymal cells of mouse liver during necrosis *in vitro* (autolysis), Lab. Invest. *11*, 986.

Tuffery, A. A. (1972), Surface extensions on BHK cells grown in monolayers and agar suspension, J. Cell Sci. *10*, 123.

Wallace, P. G., M. Huang and A. W. Linnane (1968), The biogenesis of mitochondria, J. Cell Biol. *37*, 207.

Watanabe, I., S. Donahue and N. Hoggatt (1967), Method for electron microscopic studies of circulating human leukocytes and observations on their fine structure, J. Ultrastruct. Res. *20*, 366.

Weinbaum, G., D. A. Fischman and S. Okuda (1970), Membrane modifications in nutritionally induced filamentous *Escherichia coli* B, J. Cell Biol. *45*, 493.

Winborn, W. B. and L. L. Seelig (1970), Paraformaldehyde and *s*-collidine – a fixative for preserving large blocks of tissue for electron microscopy, Tex. Rep. Biol. Med. *28*, 347.

Winters, C. and M. Slade (1971), Embedding free-floating cells in stained agar for rapid screening and subsequent ultramicrotomy, Stain technol. *46*, 161.

Woodcock, C. L. F. and P. R. Bell (1968), Features of the ultrastructure of the female gametophyte of *Myosurus minimus*, J. Ultrastruct. Res. *22*, 546.

Zeigel, R. F., G. Rabotti and M. V. A. Smith (1969), Electron microscopic observations on the association of viruses with membrane systems in hamster tumor cells propagated in tissue culture, J. natn. Cancer Inst. *43*, 653.

Zucker-Franklin, D., P. Elsbach and E. J. Simon (1971), The effect of the morphine analog levorphanol on phagocytosing leukocytes, Lab. Invest. *25*, 415.

# Chapter 4

# Dehydration

Most embedding media are not soluble in water and consequently fixed specimens are 'dehydrated' by passing them through a sequence of solutions, the last of which is miscible with the embedding medium. The two most widely used dehydrating agents are ethanol and acetone; they give very similar results with the majority of specimens and ethanol is usually preferred for general use since acetone readily takes up water and incomplete dehydration may result.

Polyester resins are not soluble in ethanol and specimens are usually dehydrated in acetone, or they are dehydrated in ethanol and then immersed in carefully dried acetone. Styrene can also be used as an intermediate solvent for polyester resins (e.g. Kurtz 1961).

Most epoxy resins are soluble in ethanol and acetone, but they mix much more readily with propylene oxide (1,2–epoxy propane). Consequently propylene oxide is frequently used as the last stage of dehydration before embedding in an epoxy resin. Xylene, toluene and styrene are also suitable as intermediate solvents for most epoxy resins, but propylene oxide is to be preferred because it is a reactive diluent and will not separate from the epoxy resin during polymerization (Luft 1961). Consequently no problems arise if a small amount of propylene oxide remains in the embedding medium. Propylene oxide should not be used if the specimen has been stained with phosphotungstic acid (PTA) during dehydration since propylene oxide extracts PTA.

## 4.1  Chemical and morphological effects of dehydration

The major chemical effect of dehydration is the extraction of lipids from the specimen by the dehydrating agents (ethanol, acetone, propylene oxide

111

etc). which are excellent lipid solvents. As much as 95% of the lipid in the specimen may be lost (Korn and Weisman 1966), the majority being extracted at concentrations of ethanol or acetone in water higher than 70% and in propylene oxide. Full details of lipid extraction during dehydration are given in the excellent review by Stein and Stein (1971). For practical purposes it can be concluded that lipid extraction can be reduced by:

(i) Fixing the specimen with osmium tetroxide before dehydration (e.g. Cope and Williams 1969a).

(ii) Fixing the specimen with uranyl acetate before dehydration (see § 2.9).

(iii) Using acetone in preference to ethanol as the dehydrating agent (e.g. Ashworth et al. 1966).

(iv) Dehydrating the specimens as rapidly as possible at low temperature.

(v) Using the partial dehydration schedule of Idelman (1964, 1965) (see § 4.2.3).

Proteins may also be extracted during dehydration. For example, Luft and Wood (1963) found that about 4% of tissue proteins were extracted during dehydration following osmium tetroxide fixation. In contrast to lipids, the proteins were extracted in the lower concentrations of ethanol in water, none being lost in 95% or 100% ethanol or in propylene oxide.

The extraction of lipid and protein during dehydration is accompanied by shrinkage of tissues, cells and subcellular components (Fig. 1.1). This extraction leads to a decrease in the diameter of sea urchin eggs (Kushida 1962), a decrease in the periodicity of nerve myelin (Moretz et al. 1969), and the shrinkage of chromosomes with an accompanying production of abrupt bends in spindle microtubules (Jensen and Bajer 1969). Some striated muscle filaments shorten by as much as 10% during dehydration in ethanol, although there is little change in dimensions during dehydration in acetone (Page and Huxley 1963). Acetone also seems to be preferable to ethanol for the dehydration of brain tissue, since specialised cell contacts in which surface membranes are closely apposed appear to be better preserved (Johnston and Roots 1967) (Fig. 4.1). However, Schultz and Karlsson (1972) consider that it is impossible to show which dehydrating agent best preserves the membrane relationships existing *in vivo*.

In lung tissue no significant change in volume is observed if the dehydration is started in 70% ethanol in water (Weibel and Knight 1964).

## 4.2   Dehydration schedules

Dehydration is accomplished by passing the fixed specimen through a graded series of solutions of increasing concentration of the dehydrating agent in water, ending with pure (or absolute) dehydrating agent. The specimens remain in the small vials in which they were fixed (Fig. 3.2) throughout dehydration. One solution is removed carefully with a Pasteur pipette and the next poured on. Care should be taken to ensure that the volume of solution used is much greater (at least 10 times) than the volume of the specimen.

For general morphological studies the effects produced during dehydration are slight compared with the changes occurring during fixation, and the schedule chosen for dehydration is not critically important. Specimens may be dehydrated at room temperature, or in the cold. If dehydration is started in the cold, the temperature is raised to room temperature when the specimens are in a 90% or 95% solution of the dehydrating agent in water.

Fixed specimens are sometimes washed in buffer before dehydration to remove excess fixative, but this is usually not necessary.

### 4.2.1 STANDARD DEHYDRATION SCHEDULE

A typical dehydration schedule for small blocks of tissue, 0.5 mm or less on each side, is:

> 50% ethanol or acetone in water    10 min
> 70% ethanol or acetone in water    10 min
> 95% ethanol or acetone in water    10 min
> 100% ethanol or acetone           15 min
> 100% ethanol or acetone           15 min

Specimens to be embedded in an epoxy resin are then passed through two changes of 15 min each of the intermediate solvent, propylene oxide, xylene or toluene. Propylene oxide must be flushed down the sink with a large volume of water after use, and care must be taken not to inhale the fumes.

These times are probably the maximum required, even for compact, well-fixed tissues. It is often unnecessary to use concentrations of dehydrating agent less than 70% and so the stage in 50% ethanol or 50% acetone can be omitted. Specimens that have to be stored overnight during dehydration are kept in 70% ethanol or 70% acetone in the cold. There is very little loss of lipid during long periods of storage in 70% ethanol (Idelman 1965).

### 4.2.2 RAPID DEHYDRATION

When required the dehydration can be made much more rapid. Dehydration times can be decreased by:

Fig. 4.1a

Fig. 4.1.   Effects of dehydration. Comparison of the appearance of closely apposed plasma membranes in rat brain after dehydration in ethanol or acetone. (a) Acetone dehydration: the plasma membranes are very closely apposed (am) or completely fused (cc). (b) Ethanol dehydration: adjacent plasma membranes (um) are separated by an extra-cellular space (es) about 10 nm wide. (m) Mitochondrion. (sv) Synaptic vesicle (from Johnston and Roots 1967).

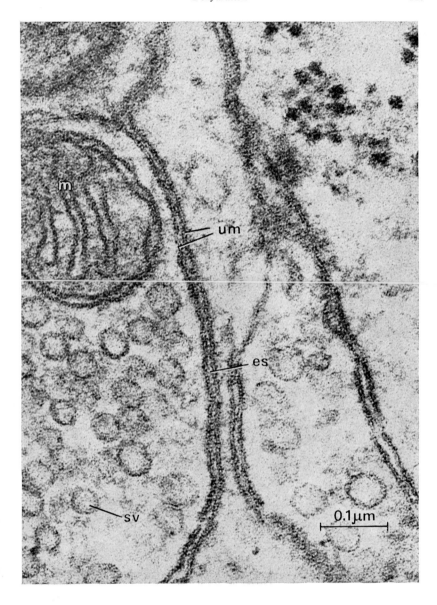

Fig. 4.1b

(i) agitating the specimens throughout dehydration

(ii) carrying out the dehydration at room temperature (and not in the cold)

(iii) using more changes of each solution

(iv) using smaller blocks of tissue.

A typical rapid dehydration schedule suggested by Coulter (1967) has proved successful with a variety of specimens. Small blocks of fixed tissue, 0.25 mm$^3$ or $2 \times 2 \times 0.1$ mm are passed through a series of solutions of increasing ethanol concentration to absolute ethanol in 2 min with constant agitation. Three changes of absolute ethanol are made in 1 min, and then three changes of propylene oxide at two minute intervals. The subsequent infiltration with embedding medium can also be done rapidly if the specimens are continuously agitated and a vacuum is applied (see § 5.3.4c).

Comparisons of rapid and standard dehydration schedules have not shown any obvious differences in the preservation of fine structure (Tormey 1965; Bencosme and Tsutsumi 1970).

4.2.3  PARTIAL DEHYDRATION

Idelman (1964, 1965) proposed a dehydration schedule which omits the use of concentrations of ethanol higher than 70% in water and of propylene oxide, thus avoiding the stages of dehydration during which the majority of the extraction of lipids occurs (Stein and Stein 1971). The dehydration schedule also depends on the fact that the epoxy resin Epon 812 (§ 5.3) is soluble in 70% ethanol in water.

*Partial dehydration schedule I*

(Idelman 1964)

| | | |
|---|---|---|
| Wash in buffer | brief | r.t. |
| 30% ethanol in water | 15 min | r.t. |
| 70% ethanol in water | 30 min | r.t. |
| 70% ethanol in water | 30 min | r.t. |
| 1:1 mixture of 70% ethanol and Epon embedding mixture (§ 5.3.1.) | 30 min | r.t. |
| and then | 30 min | 37°C |
| Epon embedding mixture | 1 hr | 37°C |
| Infiltrate and embed in Epon (§ 5.3.4a) | | |

*Key*: r.t., room temperature.

A modified version of this partial dehydration schedule has been used extensively by Stein and Stein (1971) for a great variety of lipids. They found that the loss of some lipids is reduced even further by decreasing the temperature to 0 °C.

### *Partial dehydration schedule II*

(Stein and Stein 1967)

| | | |
|---|---|---|
| 70% ethanol in water | 5 min | 0 °C |
| 70% ethanol in water | 5 min | 0 °C |
| 95% ethanol in water | 5 min | 0 °C |
| 95% ethanol in water | 5 min | 0 °C |
| Epon 812 | 1 hr | 0 °C |
| Epon 812 | 1 hr | 0 °C |
| Epon 812 | 1 hr | 0 °C |
| Epon embedding mixture (§ 5.3.1.) | overnight | 4 °C |

Infiltrate and embed in Epon

For thin pellets of cells and for monolayers of cells a rapid partial dehydration schedule can be used.

### *Rapid partial dehydration schedule*

(Robbins and Jentzsch 1967)

| | | |
|---|---|---|
| 20% ethanol–80% water | 30 sec | r.t. |
| 60% ethanol–40% Epon 812 | 5 min | r.t. |
| 30% ethanol–70% Epon 812 | 5 min | r.t. |
| 100% Epon 812 | 5 min | r.t. |

Infiltrate and embed in Epon embedding mixture

*Key:* r.t., room temperature

Robbins and Jentzsch (1967) also developed a rapid method for embedding (§ 5.3.4c)

### 4.3 Dehydrating agents

Ethanol and acetone are the standard dehydrating agents for the great majority of procedures, with the addition of propylene oxide or styrene as intermediate solvents before embedding in epoxy or polyester resins respectively. Various alternative dehydrating agents have been tested, and these are considered in this section. It is also possible to use water-soluble embedding media for dehydration as discussed in the following Chapter (§ 5.6).

4.3.1  ETHYLENE GLYCOL

Pease (1966a) investigated the possibility of dehydrating unfixed specimens using increasing concentrations of 'inert', water-soluble, permeable, organic substances as a technique for the preservation of proteins in their native state. The success of the method is thought to depend upon adequate stabilisation and consequent immobilisation of macromolecular systems before physiological damage is produced in cytomembranes. Once this immobilisation is achieved, the final stages of dehydration can be completed in a variety of ways. The method is considered to be a physical method of preservation, comparable with freeze-substitution (see § 6.3).

The simplest procedure involves the use of the dehydrating agent ethylene glycol, which affects the native conformation of proteins less than most other organic solvents (Sjöstrand and Barajas 1968). Thin slices of tissue (less than 0.5 mm thick) are isolated in a balanced salt solution containing 10% ethylene glycol, and then dehydrated as rapidly as possible to avoid excessive shrinkage. The slices are placed in a watch glass in 5 ml of salt solution containing 10% ethylene glycol and then the concentration of ethylene glycol is increased to over 50% in about 3 min by adding 7 ml of ethylene glycol drop by drop while the watch glass is agitated vigorously. At a concentration of 50% ethylene glycol the tissue becomes transparent and the subsequent dehydration schedule is not important. Dehydration is usually completed in a further 2 to 3 min. Specimens can be stored indefinitely in pure ethylene glycol at 4 °C with no obvious deterioration.

After dehydration specimens are embedded directly in hydroxypropyl methacrylate (§ 5.6.2b), or in an epoxy or polyester resin. Pease (1966a) originally suggested that Cellosolve (the monoethyl ester of ethylene glycol) be used as an intermediate solvent before embedding in an epoxy or polyester resin, but Sjöstrand and Barajas (1968) found that this resulted in complete destruction of the tissue. Consequently propylene oxide and/or toluene should be used (Pease 1973).

The main damage produced by this dehydration procedure is the extraction of lipids during dehydration and during infiltration with the embedding medium (Cope and Williams 1968, 1969b). This damage can be reduced by fixing the specimens in glutaraldehyde, or in glutaraldehyde and osmium tetroxide, in 70% ethylene glycol (Pease 1966a). Alternatively specimens are fixed briefly ($\frac{1}{2}$ to $1\frac{1}{2}$ min) with glutaraldehyde before dehydration in ethylene glycol (Sjöstrand and Barajas 1968).

After dehydration in ethylene glycol and embedding in an epoxy or polyester resin, the blocks can be sectioned relatively easily onto a trough containing ethylene glycol using a diamond knife, the whole process being anhydrous. Hydroxypropyl methacrylate blocks can only be sectioned with difficulty (Pease 1966b).

The results of this technique are of interest for comparison with conventionally-dehydrated material, but characteristic artefacts result from the tensions set up during dehydration (Pease 1973). The structure of the tissue in a surface layer about 0.1 mm thick, and of all the tissue below a depth of about 0.5 mm, is destroyed. The tissue in the narrow optimally-preserved zone shows evidence of considerable shrinkage; nuclear outlines appear scalloped and mitochondria are irregular in shape. However, much of the finer structure of the tissue is retained and some components, such as polysaccharides, are preserved better than by conventional dehydration techniques (Pease 1973).

Ethylene glycol is also useful as an intermediate solvent for specimens which are naturally dehydrated. For example, Morris (1968) fixed slices of dehydrated cysts of *Artemia salina* in dry osmium tetroxide vapours, 'washed' them in two changes of ethylene glycol and then embedded them in Epon.

### 4.3.2 POLYETHYLENE GLYCOL

Polyethylene glycol 200 (Carbowax 200) is miscible with both water and epoxy resins and is thus suitable as a dehydrating agent (Kushida 1963). Polyethylene glycol has a higher viscosity than ethanol and acetone and so longer soaking times are required. Fixed specimens are dehydrated in 50%, 70% and 90% polyethylene glycol in water for 20–40 min each, and finally in 2 changes of absolute polyethylene glycol for 40–80 min. Specimens are immersed in a 1:1 mixture of polyethylene glycol and epoxy resin embedding mixture for 1 hr before infiltration and embedding in the epoxy resin.

Polyethylene glycol is not readily soluble in polyester resins and so methyl methacrylate is used as an intermediate solvent (Kushida and Fujita 1970).

### 4.3.3 PROPYLENE OXIDE

Propylene oxide (1,2–epoxy propane) is used routinely between ethanol dehydration and infiltration with epoxy resin embedding media. Kushida (1961) showed that propylene oxide can also be used as the sole dehydrating agent for specimens to be infiltrated with Epon 812. Propylene oxide is not readily miscible with water at concentrations above 40% (Spurr 1969) and the dehydration schedule relies on the fact that water is appreciably miscible with Epon 812.

**Propylene oxide dehydration schedule**

(Kushida 1961)

| | | |
|---|---|---|
| 20% propylene oxide in water | 10–15 min | r.t. |
| 40% propylene oxide in water | 10–15 min | r.t. |
| 1:1 mixture of 70% p.o. and Epon | 10–15 min | r.t. |
| 1:1 mixture of 90% p.o. and Epon | 10–15 min | r.t. |
| 1:1 mixture of 100% p.o. and Epon | 10–15 min | r.t. |
| Epon 812 | 30 min | 40 °C |
| Epon 812 | 30 min | 40 °C |

*Key*: p.o., propylene oxide; r.t., room temperature.

The incomplete miscibility of propylene oxide with water at higher concentrations can be overcome by adding a small amount of ethanol (2.5 ml of ethanol to each 100 ml of propylene oxide). A standard dehydration schedule can then be used (Spurr 1969).

4.3.4  OTHER DEHYDRATING AGENTS

Acid alcohol was used as a dehydrating agent, following osmium tetroxide fixation, by Sandborn (1962) and he reported good preservation of matrix material and membranes. Casley-Smith (1967) found that acid alcohol did not help to preserve lipids in general, although he thought that it may improve the preservation of phospholipids.

Spurr (1969) has used hexylene glycol as a dehydrating agent with the aim of helping to avoid loss of lipids.

Watson (1962) used a mixture of acrolein (§ 2.6.1) and ethanol as a dehydrating agent for the preservation of proteins, while Robison and Lipton (1969) used acrolein alone, following fixation in a mixture of acrolein and potassium dichromate (§ 2.8.6). This procedure revealed subunit detail in microtubules which appeared 'negatively' stained.

REFERENCES

Ashworth, C. T., J. S. Leonard, E. H. Eigenbrodt and F. J. Wrightman (1966), Hepatic intracellular osmiophilic droplets. Effect of lipid solvents during tissue preparation, J. Cell Biol. *31*, 301.
Bencosme, S. A. and V. Tsutsumi (1970), A fast method for processing biologic material for electron microscopy, Lab. Invest. *23*, 447.

Casley-Smith, J. R. (1967), Some observations on the electron microscopy of lipids, Jl R. microsc. Soc. *87*, 463.

Cope, G. H. and M. A. Williams (1968), Quantitative studies on neutral lipid preservation in electron microscopy, Jl R. microsc. Soc. *88*, 259.

Cope, G. H. and M. A. Williams (1969a), Quantitative studies on the preservation of choline and ethanolamine phosphatides during tissue preparation for electron microscopy. I. Glutaraldehyde, osmium tetroxide, Araldite methods, J. Microscopy *90*, 31.

Cope, G. H. and M. A. Williams (1969b), Quantitative studies on the preservation of choline and ethanolamine phosphatides during tissue preparation for electron microscopy. II. Other preparative methods, J. Microscopy *90*, 47.

Coulter, H. D. (1967), Rapid and improved methods for embedding biological tissues in Epon 812 and Araldite 502, J. Ultrastruct. Res. *20*, 346.

Idelman, S. (1964), Modification de la technique de Luft en vue de la conservation des lipides en microscopie électronique, J. Microscopie *3*, 715.

Idelman, S. (1965), Conservation des lipides par les techniques utilisées en microscopie électronique, Histochemie *5*, 18.

Jensen, C. and A. Bajer (1969), Effects of dehydration on the microtubules of the mitotic spindle, J. Ultrastruct. Res. *26*, 367.

Johnston, P. V. and B. I. Roots (1967), Fixation of the central nervous system by perfusion with aldehydes and its effect on the extracellular space as seen by electron microscopy, J. Cell Sci. *2*, 377.

Korn, E. D. and R. A. Weisman (1966), Loss of lipids during preparation of amoebae for electron microscopy, Biochim. biophys. Acta *116*, 309.

Kurtz, S. M. (1961), A new method for embedding tissues in Vestopal W, J. Ultrastruct. Res. *5*, 468.

Kushida, H. (1961), Propylene oxide as a dehydrating agent for embedding with epoxy resins, J. Electron Microscopy, *10*, 203.

Kushida, H. (1962), A study of cellular swelling and shrinkage during fixation, dehydration and embedding in various standard media, J. Electron Microscopy, *11*, 135.

Kushida, H. (1963), An improved epoxy resin 'Epok 533', and polyethylene glycol 200 as a dehydrating agent, J. Electron Microscopy, *12*, 167.

Kushida, H. and K. Fujita (1970), Polyethylene glycol 200 as a dehydrating agent for embedding with polyester resins, J. Electron Microscopy, *19*, 391.

Luft, J. H. (1961), Improvements in epoxy resin embedding methods, J. biophys. biochem. Cytol. *9*, 409.

Luft, J. H. and R. L. Wood (1963), The extraction of tissue protein during and after fixation with osmium tetroxide in various buffer systems, J. Cell Biol. *19*, 46A.

Moretz, R. C., C. K. Akers and D. F. Parsons (1969), Use of small angle X-ray diffraction to investigate disordering of membranes during preparation for electron microscopy. I. Osmium tetroxide and potassium permanganate, Biochem. biophys. Acta *193*, 1.

Morris, J. E. (1968), Dehydrated cysts of *Artemia salina* prepared for electron microscopy by totally anhydrous techniques, J. Ultrastruct. Res. *25*, 64.

Page, S. G. and H. E. Huxley (1963), Filament lengths in striated muscle, J. Cell Biol. *19*, 369.

Pease, D. C. (1966a), The preservation of unfixed cytological detail by dehydration with 'inert' agents, J. Ultrastruct. Res. *14*, 356.

Pease, D. C. (1966b), Anhydrous ultrathin sectioning and staining for electron microscopy, J. Ultrastruct. Res. *14*, 379.

Pease, D. C. (1973), Substitution techniques, in: Advanced techniques in biological electron microscopy, J. K. Koehler, ed. (Springer-Verlag, Berlin). p. 35.

Robbins, E. and G. Jentzsch (1967), Rapid embedding of cell culture monolayers and suspensions for electron microscopy, J. Histochem. Cytochem. *15*, 181.

Robison, W. G. and B. H. Lipton (1969), Advantages of dichromate-acrolein fixation for preservation of ultrastructural detail, J. Cell Biol. *43*, 117a.

Sandborn, E. (1962), A method of dehydration for improved visualization of lipids, membranes and other cytoplasmic inclusions in tissues to be embedded in Epon, Proc. 5th Int. Congr. Electron Microscopy, Philadelphia *2*, P-12.

Schultz, R. L. and U. L. Karlsson (1972), Brain extracellular space and membrane morphology variations with preparative procedures, J. Cell Sci. *10*, 181.

Sjöstrand, F. S. and L. Barajas (1968), Effect of modifications in conformation of protein molecules on structure of mitochondrial membranes, J. Ultrastruct. Res. *25*, 121.

Spurr, A. R. (1969), A low-viscosity epoxy resin embedding medium for electron microscopy, J. Ultrastruct. Res. *26*, 31.

Stein, O. and Y. Stein (1967), Lipid synthesis, intracellular transport, storage, and secretion. 1. Electron microscope radioautographic study of liver after injection of tritiated palmitate or glycerol in fasted and ethanol-treated rats, J. Cell Biol. *33*, 319.

Stein, O. and Y. Stein (1971), Light and electron microscopic radioautography of lipids: techniques and biological applications, Adv. Lipid Res. *9*, 1.

Tormey, J. McD. (1965), Artifactual localization of ferritin in the ciliary epithelium *in vitro*, J. Cell Biol. *25*, 1.

Watson, M. L. (1962), Considerations of nucleic acid morphology in fixed tissues, Proc. 5th Int. Congr. Electron Microscopy, Philadelphia *2*, 0-5.

Weibel, E. R. and B. W. Knight (1964), A morphometric study on the thickness of the pulmonary air–blood barrier, J. Cell Biol. *21*, 367.

# Chapter 5

# Embedding

In the final stage of preparing a biological specimen in a form suitable for thin-sectioning it is infiltrated with a liquid embedding medium which is then polymerized to produce a solid block.

## 5.1 Embedding media for electron microscopy

Ideally an embedding medium should have the following properties:
  (i) Easy availability
  (ii) Uniformity from one batch to another
  (iii) Solubility in dehydrating agents
  (iv) Low viscosity as a monomer
  (v) Polymerize uniformly
  (vi) Little volume change on polymerization
  (vii) Ease of sectioning
  (viii) Stability in the electron beam

Three main types of embedding media are in general use; the epoxy resins, the polyester resins and the methacrylates. None of the embedding media so far developed possesses all the desired qualities, and some of the properties may be incompatible with others. For example, a medium which polymerizes with little shrinkage usually has a high initial viscosity. The embedding media widely used in electron microscopy are listed in Table 5.1, and their properties are discussed in the following sections of this chapter. The epoxy resins (§ 5.3) are the most widely used since they have most of the properties required. The polyester resins (§ 5.4) have similar excellent properties, but are less easily available and some of the components of the embedding medium are not stable during storage. The methacrylates (§ 5.5) polymerize unevenly and are unstable under electron bombardment. Consequently, they are not suitable for general use, although the water-soluble methacrylates (§ 5.6.2) are valuable as embedding media in cytochemical studies.

TABLE 5.1

Embedding media for electron microscopy

| Manufacturer | Trade name | Introduced by: |
|---|---|---|
| *Epoxy Resins* | | |
| Shell | Epon 812 (originally 562) and Epon 815 | Kushida (1959) |
| | Epon 812 (named Epikote 812 in Europe) | Gibbons (1959); Finck (1960); Luft (1961) |
| Ciba | Araldite CY 212 (originally Araldite M) (sold by Fluka and others as Durcupan ACM) | Glauert et al. (1956); Glauert and Glauert (1958) |
| | Araldite 6005 | Mollenhauer (1959) |
| | X133/2097 (sold by Fluka as Durcupan) | Stäubli (1960) |
| | Araldite 502 | Finck (1960); Luft (1961) |
| | Araldite 506 | Mollenhauer (1964) |
| Marblette | Maraglas 655 | Freeman and Spurlock (1962); Spurlock et al. (1963) |
| Oken Shoji | Epok 533 | Kushida (1963b) |
| Dow | DER 332 | Lockwood (1964) |
| | DER 334 | Winborn (1965) |
| Union Carbide | ERL 4206 | Spurr (1969) |
| *Polyester resins* | | |
| Martin Jaeger | Vestopal W | Ryter and Kellenberger (1958a, b). |
| Riken Goseijushi | Rigolac 2004 and Rigolac 70F | Kushida (1960a) |
| Pittsburg Plate Glass | Selectron 5003 and Selectron 5214 | Low and Clevenger (1962) |
| Société des Usines Chimıques Rhône-Poulenc | Rhodester 1108 | Argagnon and Enjalbert (1964) |
| B.I.P. Chemicals | Beetle 4116 and Beetle 4134 | Rampley and Morris (1972) |
| *Methacrylates* | | |
| Various | *n*-butyl methacrylate and methyl methacrylate | Newman et al. (1949) |
| Various | glycol methacrylate | Rosenberg et al. (1960) |
| Various | hydroxypropyl methacrylate | Leduc and Holt (1965) |

## 5.1.1 CHEMICAL AND MORPHOLOGICAL EFFECTS OF EMBEDDING MEDIA

The changes occurring in the specimen during embedding are slight compared with those occurring during fixation (Chapter 2) and dehydration (Chapter 4). There is little dimensional change during infiltration of the embedding medium into most specimens (Fig. 1.1; Kushida 1962b; Moretz et al. 1969a), although isolated cells, such as amoebae, may shrink in viscous embedding media (Szubinska 1971). There is also little shrinkage (usually less than 2%) during the polymerization of epoxy and polyester resins (Fig. 1.1). Large dimensional changes result from embedding in methacrylates, including three-dimensional methacrylate (Fig. 1.1). Epoxy resin monomers extract small amounts of lipid, the amount depending on the preceding fixation and dehydration schedule (Korn and Weisman 1966; Casley-Smith 1967; Cope and Williams 1968; 1969a, b; Stein and Stein 1971). The methacrylates are powerful lipid solvents, even at low temperatures (Cope and Williams 1968; 1969b; Stein and Stein 1971).

## 5.2 Standard embedding methods

Embedding media should be handled with great care. Many of the components can cause dermatitis if they come into contact with the skin, and consequently all procedures should be carried out as cleanly as possible. In addition, the vapours of the components are often toxic and particular care should be taken when handling embedding media of low viscosity which tend to be more volatile. Embedding media should be handled in a fume cupboard whenever possible.

The standard method for embedding specimens in gelatin or polyethylene capsules is similar for all types of embedding medium. After fixation (Chapter 3) and dehydration (Chapter 4) the specimens will be soaking in the final dehydrating solution in small vials (Fig. 3.2) and are ready for infiltration with the embedding medium. Complete embedding mixtures, containing the accelerator or activator, are usually used for infiltration and are prepared freshly during the later stages of dehydration.

The specimens are infiltrated with the embedding medium by passing them through a sequence of solutions until the dehydrating agent has been completely replaced by the final embedding mixture. The different schedules required for the various embedding media are described in the later sections of this chapter. In a typical schedule, the dehydrating agent (or intermediate solvent) is removed with a pipette and replaced with a solution consisting of equal parts of the dehydrating agent and the embedding medium. After 1 hr this solution is replaced with pure embedding medium and the vials are left overnight with their tops off to allow any remaining dehydrating agent to evaporate. It is difficult to pour viscous embedding media and they are best transferred in a wide-mouthed pipette.

The infiltration of specimens with viscous embedding media can be speeded up by placing the vials on a slow rotary shaker. Jurand and Ireland (1965) and Kushida (1969a) have designed rotary mixers specially for embedding (Figs. 5.1 and 5.2), which are available commercially (see Appendix). These mixers can also be used to shorten fixation and dehydration schedules, and in the preparation of fixatives and embedding mixtures (Fig. 5.3).

As an alternative to using a shaker, a continuous stream of gas bubbles can be used to keep the embedding medium circulating around the specimens (Kushida and Fujita 1971). Nitrogen is used for methacrylates, styrene and polyester resins (since oxygen inhibits their polymerization), and dehydrated air for epoxy resins.

Fig. 5.1.   Cross-sectional diagram of the rotary shaker designed by Kushida (1969a): (A) vial; (B) metal plate; (C) pan; (D) spherical bearing; (E) shaft; (F) self-aligning bearing; (G) carriage (H) pulley; (I) belt; (J) induction motor.

Fig. 5.2. A general view of the shaker designed by Kushida (1969a).

Delicate tissue specimens may be adversely affected by sudden changes in viscosity and it is therefore advisable to increase the concentration of the embedding medium in a continuous process. Steinbrecht and Ernst (1967) describe a method of continuous infiltration in which the vials containing the specimens in propylene oxide are placed on a rotating mixer similar to the device designed by Jurand and Ireland (1965) and the embedding medium (an epoxy resin) is added dropwise.

In the standard embedding method, gelatin or polyethylene capsules are used as embedding moulds to provide blocks of a convenient size for thin sectioning. Gelatin capsules are obtainable in a range of sizes (No. 00, 0, 1, 2, etc.) (See Appendix); size No. 0 is convenient for general use. Polyethylene capsules (Fig. 5.4) of various designs are available commercially and are known as BEEM or TAAB capsules, depending on the supplier (see Appendix). These capsules have pyramid-shaped ends and yield pre-formed blocks that require the minimum of trimming before sectioning (Reid 1974).

Fig. 5.3.   The shaker designed by Kushida (1969a) adapted to hold beakers and flasks
for the preparation of fixatives and embedding media.

Taab also supply capsules made of polypropylene for polymerization at
high temperatures.

The capsules must be completely dry before use and are best stored in a
60 °C oven overnight, or longer, before use. A set of capsules is placed in a
suitable support, such as the lid of a cardboard box punched with an array of
holes (Fig. 5.5) or special trays which are available commercially (see
Appendix). A specimen is removed from the vial with fine tweezers or a
pipette and placed at the bottom of the capsule. The capsule is filled about
half full with the final embedding mixture and the specimen is guided back
to the exact bottom of the capsule, if necessary. A label indicating the experiment and sample number (use a typewriter or Indian ink) is placed in each
capsule in such a position that the number is visible from outside the capsule
(Fig. 5.5).

The embedding medium is polymerized by heating the capsules in an oven or by irradiation with ultra-violet (UV) light. Polymerization by UV irradiation has the advantage that it can be carried out at room temperature or in the cold. For example, Shinagawa and Uchida (1961) polymerized methacrylate by UV light in a cold room and found that the temperature of the specimens did not rise above 24 °C.

An ordinary sun lamp is adequate as a source of UV light (e.g. Leduc and Holt 1965). The capsules are placed a few inches away from the lamp.

Fig. 5.4.   Various designs of BEEM polyethylene capsules designed to produce pre-shaped blocks.

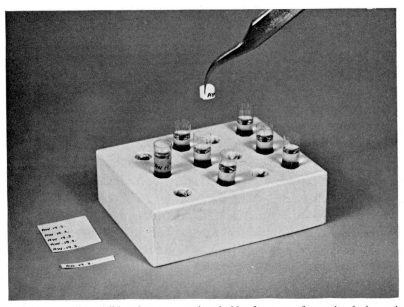

Fig. 5.5.   A cardboard lid makes a convenient holder for a set of capsules during polymerization of the embedding medium. A label is about to be inserted into one of the gelatin capsules.

Alternatively, a special apparatus can be constructed, such as that designed by Shinagawa et al. (1962) in which 5 UV lamps are attached to the inside of a metal arch. Cole (1968) described a simple and cheap apparatus which is easily constructed in the laboratory. Polymerization was carried out at 4 °C, 30 °C or 42 °C by placing the apparatus in a freezer at − 16 °C, or in a constant temperature room at 27 °C or in an incubator at 35 °C respectively. The UV source was a General Electric black-light lamp. A more complex apparatus is described by McGee-Russell and De Bruijn (1964). Suitable UV lamps are available commercially (see Appendix).

## 5.3   Epoxy resins

The epoxy resins were tested as embedding media for electron microscopy (Maaløe and Birch-Andersen 1956; Glauert et al. 1956) in a search for an alternative to the methacrylates (§5.5). In common with polyester resins (§5.4), epoxy resins have the advantages that they polymerize uniformly with very little change of volume and are relatively stable in the electron beam (Reid 1974). Their main disadvantage is their high viscosity which necessitates a lengthy infiltration procedure as compared with the methacrylates.

Chemically the epoxy resins are polyaryl ethers of glycerol bearing terminal epoxy groups. They range from viscous liquids to fusible solids and may be polymerized by a variety of bifunctional setting agents which add across the epoxy groups of the resin molecules to give three-dimensional structures. These setting agents are commonly aliphatic polyamines or aromatic anhydrides, but the number of possible setting agents is very large and this enables a wide range of products with very different characteristics to be obtained. The polymerization can be made to occur with a volume shrinkage as low as 2%, a comparable figure for methacrylate being 15 to 20%. This relatively small change in volume is due, on the one hand, to the polymerization being an addition process, and, on the other, to the fact that the resin is highly associated in the uncured state (Glauert and Glauert 1958).

The uniformity of the polymerization of epoxy resins is due to the fact that the hardening is an addition reaction and is not dependent on an initiator. In consequence the reaction is not affected by the presence of impurities, unlike the methacrylates in which components of the specimen itself may initiate polymerization (§5.5).

### 5.3.1   EPOXY RESIN EMBEDDING MEDIA

The first epoxy resin embedding medium to be tested (Maaløe and Birch-

Andersen 1956) consisted of a Shell epoxy resin (designated EPO) and an aliphatic polyamine hardener (diethylene triamine). This hardener is very viscous and the mixture sets rapidly even at room temperature. The mixture also has a limited compatibility with ethanol and it is very difficult to get adequate penetration of the embedding medium into the specimen. Subsequently Glauert et al. (1956) tested a series of Ciba epoxy resins (Araldites) and an anhydride hardener (dodecenyl succinic anhydride, DDSA). The reaction between the epoxy resin and this hardener is much slower than in an EPO mixture and the hardener is less viscous than diethylene triamine. In addition, the mixture of Araldite and DDSA is readily soluble in ethanol so that good penetration of the embedding medium into the specimen is easily achieved. The reaction between the epoxy resin and DDSA is slow and an amine accelerator is added to hasten polymerization.

The Araldites (502, 506 and 6005) manufactured in the USA are not identical to those made in England (CY 212) and some difficulties were experienced in using them as embedding media (Richardson et al. 1960). Consequently Finck (1960) and Luft (1961) investigated the suitability of Shell epoxy resins (Epons, named Epikotes in Europe) for electron microscopy, and showed that embedding media based on Epon 812 are excellent for the purpose. At the same time Kushida (1959) in Japan developed an embedding medium containing a mixture of Epon 812 and Epon 815, and later he used an epoxy resin (Epok) manufactured in Japan by Oken Shoji (Kushida 1963b). Embedding media based on epoxy resins manufactured by Marblette (Maraglas), Dow (DER) and Union Carbide (ERL) have also been successfully used for electron microscopy.

Most of the epoxy resin embedding media have the same basic components as the original Araldite mixture, consisting of an epoxy resin, a hardener and an accelerator, sometimes with the addition of a plasticizer or epoxy flexibilizer. These embedding media are listed in Table 5.2. It will be noted that some media contain more than one epoxy resin (sometimes from more than one manufacturer) and that others contain more than one hardener. In some mixtures a reactive epoxy flexibilizer (Cardolite NC-513, DER 732 or DER 736) acts as hardener as well as plasticizer. The final blocks range from crystal clear (e.g. Maraglas 655) to light gold (e.g. Araldite) in colour.

All the embedding media listed in Table 5.2 are suitable for electron microscopy and a decision on which to use is usually based on availability and the uniformity of the epoxy resin from one batch to another. There are reports that Maraglas 655 (Erlandson 1964) and DER 334 (Winborn 1965) show some variability and that it is desirable to run blanks on each new can.

TABLE 5.2

Epoxy resin embedding media

| Epoxy resin | | Hardener | | Accelerator | | Additives | | Features | Reference |
|---|---|---|---|---|---|---|---|---|---|
| Araldite CY 212 | 10 ml | DDSA | 10 ml | DMP-30 | 0.5 ml | DBP | 1 ml | First use of Araldite | Glauert and Glauert (1958) |
| Araldite CY 212 | 10 ml | DDSA | 10 ml | DMP-30 | 0.5 ml | | | Plasticizer omitted | Richardson et al. (1960) |
| Araldite CY 212 | 10 ml | DDSA 9 ml + MNA | 1 ml | DMP-30 | 0.4 ml | | | For harder specimens | Robson (1964) |
| Araldite CY 212 | 10 ml | DDSA | 10 ml | BDMA | 0.4 ml | | | Standard 1974 | Parker (1972) |
| Araldite 502 | 10 ml | DDSA | 10 ml | BDMA | 0.2 ml | | | U.S. equivalents of original Araldite mixture | Finck (1960) |
| Araldite 502 | 27 ml | DDSA | 23 ml | DMP-30 | 0.75–1.0 ml | | | | Luft (1961) |
| Araldite 502 | 68 ml | DDSA | 19 ml | DMP-30 | 3 ml | TAC | 10 ml | TAC added to improve sectioning properties | Winborn (1965) (Table 5.10) |
| Araldite 502 | 10 ml | DDSA | 7.8 ml | DMP-30 | 1.5% | | | Anhydride/epoxy ratio adjusted | Coulter (1967) (Table 5.9) |
| Araldite 6005 | 10 ml | DDSA | 7 ml | DMP-30 | 0.24 ml | | | U.S. equivalents of original Araldite mixture | Richardson et al. (1960) |
| Araldite 6005 | 10 ml | DDSA | 10 ml | BDMA | 0.2–0.3 ml | DBP | 1 ml | | Parsons and Darden (1961) |
| Araldite 506 | 50 ml | | | BDMA | 3% | Cardolite NC-513 + DBP | 25 ml 1–2 ml | DDSA replaced with reactive epoxy flexibilizer | Mollenhauer (1964) (Table 5.8) |
| Epon 812 | 10 ml | DDSA | 19 ml | BDMA | 1% | | | First use of Epon 812 in the U.S. | Finck (1960) |
| Epon 812 | 10 ml | DDSA 18 ml + | | BDMA | 1% | | | Add HHPA for | Finck (1960) |

| Resin | Hardener | Accelerator | Flexibilizer/Plasticizer | Comments | Reference |
|---|---|---|---|---|---|
| Epon 812 | DDSA + MNA | DMP-30 1.5–2.0% | NC-513 | reactive flexibilizer Standard 1974 DDSA/MNA ratio determines hardness | Luft (1961) (Table 5.7) |
| Epon 812 | DDSA + MNA | DMP-30 1.5% | | Anhydride/epoxy ratio adjusted to improve sectioning properties | Coulter (1967) (Table 5.9) |
| Epon 812 | DDSA + MNA | DMP-30 or BDMA | | ⎱ | Burke and Geiselman (1971) |
| Epon 812 10 ml | MNA 8.5 ml | DMP-30 0.15–0.3 ml | Thiokol LP-8 1.5–3.0 ml | Embedding at room temp. | Kushida (1965c) |
| Epon 812 8 ml | MNA 8.9 ml | DMP-30 1.5–2.0% | DER 732 2 ml | DDSA replaced with a reactive flexibilizer | Kushida (1966b) |
| Epon 812 2 ml | MNA 8.9 ml | DMP-30 1.5–2.0% | DER 736 8 ml | ⎰ | Kushida (1967) |
| Epon 812 10 ml | NSA 13 ml | DMP-30 1.5–2.0% | | Low viscosity hardener | Kushida (1971) |
| Epon 812 + Epon 815 10 ml | DDSA 14–16 ml | DMP-30 2 ml or BDMA | | First use of Epon 812 | Kushida (1959) |
| Epon 812 + Epon 815 | DDSA | DMP-30 | | Epon 812/Epon 815 ratio determines hardness | Shinagawa et al. (1962) (Table 5.6) |
| Araldite 502 + Epon 812 20 ml 25 ml | DDSA 60 ml | DMP-30 2.5 ml | DBP 2–4 ml | First use of a mixture of Araldite and Epon | Voelz and Dworkin (1962) |
| Araldite CY 212 (or Araldite 502 or 6005) + Epon 812 15 ml 25 ml | DDSA 55 ml | DMP-30 1.5% or BDMA 3% | | See comments in text | Mollenhauer (1964) (Table 5.8) |
| Araldite 506 + Epon 812 81 ml 62 ml | DDSA 100 ml | DMP-30 1.5% or BDMA 3% | DBP 4–7 ml | Popular 1974 | Mollenhauer (1964) |

*Table 5.2 (continued)*

| Epoxy resin | Hardener | Accelerator | Additives | Features | Reference |
|---|---|---|---|---|---|
| Araldite 502 20 ml + Epon 812 25 ml | DDSA 60 ml | DMP-30 1.5–2.5% | | Sections susceptible to action of hydrolases | Anderson and André (1968) |
| Araldite 502 35 ml + Epon 812 40 ml | DDSA 40 ml | DMP-30 1.5–2.5% | | | |
| Epok 533 10 ml | MNA 8.6 ml | DMP-30 1.5–2.0% | Carbowax 200 0–2% | Epoxy resin manufactured in Japan. Low viscosity | Kushida (1963b) |
| Maraglas 655 | | BDMA 2% | Cardolite NC-513 + DBP | First use of Maraglas. Hardness varied with content of flexibilizer and plasticizer | Freeman and Spurlock (1962) (Table 5.4) |
| Maraglas 655 68 ml | | BDMA 2 ml | Cardolite 20 ml NC-513 + DBP 10 ml | Modified procedure | Spurlock et al. (1963) |
| Maraglas 655 36 ml | | BDMA 1 ml. | DER 732 8 ml + DBP 5 ml | Cardolite NC-513 replaced with DER 732 | Erlandson (1964) |
| Maraglas 655 48 ml | | DMP-30 2 ml | Cardolite 40 ml NC-513 + TAC 10 ml | TAC added to improve sectioning properties | Winborn (1965) (Table 5.10) |
| DER 332 | DDSA | DMP-30 | DER 732 | First use of Dow epoxy resin. Hardness adjusted with content of flexibilizer | Lockwood (1964) (Table 5.5) |
| DER 332 7 g | DDSA 8 g | DMP-30 0.3 g | DER 732 3.2 g | Modification for amphibian eggs | Bluemink (1970) |
| DER 334 48 ml | DDSA 40 ml | DMP-30 2 ml | TAC 10 ml | TAC added to improve sectioning properties | Winborn (1965) (Table 5.10) |
| ERL 4206 10 g | NSA 26 g | DMAE 0.4 g | DER 736 6 g | Low viscosity | Spurr (1969) (Table 5.3) |

Key to Table 5.2

*Epoxy resins*

| | |
|---|---|
| Araldite CY 212 | Manufactured by Ciba in Europe. Previously named Araldite M. Sold by Fluka (see Appendix) as Durcupan ACM. Viscosity 1300–1650 cps at 25 °C. |
| Araldite 502,6005 and 506 | Manufactured by Ciba in the USA. |
| Epon 812 (Epikote 812 in Europe) | A glycerol-based aliphatic epoxy resin. Manufactured by Shell. Readily takes up water (hygroscopic). Viscosity 150–210 cps at 25 °C. |
| Epon 815 | Manufactured by Shell. |
| Epok 533 | Manufactured by Oken Shoji. Low viscosity. |
| Maraglas 655 | Manufactured by Marblette. Viscosity 500 cps at 25 °C. Variability from can to can reported. |
| DER 332 | Manufactured by Dow. A very pure epoxide. |
| DER 334 | Manufactured by Dow. Polymerization time varies with batch of resin. Viscosity 500–700 cps. at 25 °C. |
| ERL 4206 | Vinyl cyclohexene dioxide. Manufactured by Union Carbide. Viscosity 7.8 cps at 25 °C. |

*Hardeners*

| | |
|---|---|
| DDSA | dodecenyl succinic anhydride. Viscosity 290 cps at 25 °C. Trade name HY 964 (previously 964B) (Ciba). |
| HHPA | hexahydrophthalic anhydride. A white crystalline solid. Melting point 35 °C. Trade name Epikure HPA (Shell) |
| MNA or NMA | methyl nadic anhydride or nadic methyl anhydride. Viscosity 175–275 cps at 25 °C. Reacts with permanganates (Reedy 1965). |
| NSA | nonenyl succinic anhydride. Viscosity 117 cps at 25 °C. |

*Accelerators*

| | |
|---|---|
| DMP-30 | 2,4,6-tridimethylamino methyl phenol. Trade names DMP-30 and DY 064 (previously 964C) (Ciba). |
| BDMA | benzyl dimethylamine. Trade name DY 062 (Ciba). |
| DMAE | dimethylaminoethanol. Gives longer useful pot-life than DMP-30 or BDMA. Trade name S-1 (Pennsalt). |

*Additives*

| | |
|---|---|
| DBP | dibutyl phthalate. Plasticizer. |
| TAC | triallyl cyanurate |
| Cardolite NC-513 | A long chain mono-epoxide resin flexibilizer. One epoxy group per molecule. Compatible with propylene oxide. Viscosity 50 cps at 25 °C. |
| Thiokol LP-8 | A liquid polythiodithiol polymer of low viscosity. A reactive flexibilizer. |
| DER 732 | A polyglycol diepoxide flexibilizer. Manufactured by Dow. Viscosity 55–100 cps at 25 °C. |
| DER 736 | Digycidyl ether of polypropylene glycol. Manufactured by Dow. Shorter chain length than DER 732. Viscosity 30–60 cps at 25 °C. |
| Carbowax 200 | polyethylene glycol 200 |

This is tedious and time consuming. In the majority of laboratories a standard embedding medium and procedure is adopted and then used for many years without modification for a great range of specimens.

A survey of the recent literature indicates that the Epon 812 mixture of Luft (1961) is a standard embedding medium throughout the world, while Araldite CY 212 (Glauert et al. 1956) is widely used in Europe where it is easily obtainable. The mixture of Epon 812 and Araldite CY 212 or 506 proposed by Mollenhauer (1964) is also popular, and the low-viscosity embedding medium of Spurr (1969) based on ERL 4206 is valuable for specimens that are hard to infiltrate with more viscous epoxy resins.

5.3.2 CHARACTERISTICS OF EPOXY RESIN EMBEDDING MEDIA

5.3.2.a   *Viscosity*
The main differences between the various epoxy resins is in their viscosity. They can be placed in the following order, ranging from high to low viscosity:

> Araldite 502 (3000)
> Araldite CY 212 (1300–1650)
> DER 334 (500–700)
> Maraglas 655 (500)
> Epon 812 (150–210)
> ERL 4206 (7.8)

The viscosity in centipoise (cps) at 25 °C is given in brackets.

Longer infiltration times are required with the more viscous media (§ 5.3.4), and for very dense specimens it is advisable to choose a medium of low viscosity.

The viscosity of the complete embedding medium is also influenced by the viscosity of the hardener, and can be reduced by using NSA (117 cps) in preference to HHPA (which is solid at 25 °C), DDSA (290 cps) or MNA (175–275 cps) (Spurr 1969; Kushida 1971). The viscosity is also less for media containing plasticizers or flexibilizers added to modify the hardness and sectioning properties of the final block.

The viscosity of an epoxy resin embedding medium usually increases on the addition of the accelerator, and then continues to increase with time at a rate depending on the temperature as polymerization proceeds. For example, Araldite mixtures become too viscous to use after about 48 hr at room temperature.

Spurr (1969) selected the epoxy resin ERL 4206 (7.8 cps), the hardener

NSA (117 cps) and the reactive flexibilizer DER 736 (30–60 cps), all of which have a low viscosity, for his low-viscosity embedding medium. The complete medium has a viscosity of 60 cps at 25 °C. This medium must *always* be used in a fume cupboard; the volatile vapours may be harmful.

TABLE 5.3
Low-viscosity epoxy resin embedding medium (Spurr 1969)

| Mixture | Standard medium | Modifications | | | |
|---|---|---|---|---|---|
| | A | B | C | D | E |
| | (ml) | (ml) | (ml) | (ml) | (ml) |
| ERL 4206 | 10.0 | 10.0 | 10.0 | 10.0 | 10.0 |
| DER 736 | 6.0 | 4.0 | 8.0 | 6.0 | 6.0 |
| NSA | 26.0 | 26.0 | 26.0 | 26.0 | 26.0 |
| DMAE (S 1) | 0.4 | 0.4 | 0.4 | 1.0 | 0.2 |
| *Hardness* | Firm | Hard | Soft | | |
| *Polymerization* time at 70°C (hr) | 8 | 8 | 8 | 3 | 16 |
| *Pot life* (days) | 3–4 | | 3–4 | 3–4 | 2 | 7 |

The rate of polymerization and viscosity of the medium, and the hardness of the final block, can be modified by adjusting the relative amounts of ERL 4206, DER 736 and the accelerator dimethylaminoethanol (DMAE) (Table 5.3). Modifications favouring one feature generally involve sacrifice of other qualities of the medium. For example, a medium with a rapid rate of polymerization (mixture D) has a short pot life, while a medium with a maximum pot life and minimum viscosity (mixture E) requires a longer polymerization time.

5.3.2.b  *Hardness of the final block*
The hardness of the final block is influenced by each of the components of the embedding medium and can be modified in a number of ways.

(i) *Plasticizers and flexibilizers.* The simplest way to reduce the hardness of the final block is to add a plasticizer, dibutyl phthalate (DBP) (Glauert et al. 1956) or a flexibilizer, Cardolite NC-513 (Finck 1960), DER 732 (Lockwood 1964), Thiokol LP-8 (Kushida 1965c) or DER 736 (Kushida 1967). The flexibilizers are usually to be preferred to the plasticizers since they react with the epoxy resin and become part of the cross-linked structure. Consequently they are less likely to be lost under electron bombardment. In some embedding media, however, dibutyl phthalate is necessary to prevent the block becoming too brittle (e.g. Erlandson 1964) and to improve the sectioning properties (Mollenhauer 1964).

TABLE 5.4

Variation of hardness with the content of plasticizer and flexibilizer
(Freeman and Spurlock 1962)

| Mixture | A | B | C | D | E |
|---------|---|---|---|---|---|
| | (%) | (%) | (%) | (%) | (%) |
| Maraglas 655 | 60 | 60 | 65 | 65 | 70 |
| Cardolite NC-513 | 40 | 30 | 30 | 20 | 20 |
| DBP | – | 10 | 5 | 15 | 10 |
| | soft ─────────────────────────→ hard | | | | |

All mixtures contain 2% BDMA as accelerator.

Mixture E is generally preferred so long as care is taken to obtain adequate infiltration of the embedding medium into the specimen (Spurlock et al. 1963).

Erlandson (1964) reported that the mixture recommended by Freeman and Spurlock (1962) (Table 5.4) yielded quite brittle blocks and suggested replacing Cardolite NC-513 with the polyglycol diepoxide flexibilizer DER 732. He recommends a mixture containing Maraglas 655 (72%), DER 732 (16%), DBP (10%) and BDMA (2%). The sectioning properties of the blocks can be modified by varying the proportions of Maraglas 655 and DER 732. The hardness of the blocks decreases with increasing concentrations of DER 732; for harder blocks as little as 10% of DER 732 may be used, while softer blocks are obtained by increasing the concentration to 30%.

Lockwood (1964) tested a large group of mixtures of the epoxy resin DER 332 in combination with the reactive flexibilizer DER 732 and recommended the three mixtures listed in Table 5.5.

TABLE 5.5

Variation of hardness with the content of
reactive flexibilizer (Lockwood 1964)

| Mixture | 1 | 2 | 3 |
|---------|---|---|---|
| | (ml) | (ml) | (ml) |
| DER 332 | 7 | 7 | 6 |
| DDSA | 5 | 5 | 10 |
| DER 732 | 3 | 2 | 3 |
| DMP–30 | 0.30 | 0.28 | 0.38 |
| *Hardness* | soft | hard | soft |

Mixture No. 1 gives soft blocks and is particularly suitable for soft tissues (kidney, liver, etc.) and for isolated cells. Mixture No. 2 gives harder blocks and is preferable for tissues rich in collagen. Mixture No. 3 gives blocks with the same hardness as mixture No. 1 but gives stronger sections due to the greater concentration of DDSA. Bluemink (1970) modified the DER 332 embedding medium proposed by Lockwood (1964) in accordance with the recommendations of Coulter (1967) concerning anhydride/epoxy ratios (§5.3.2c), and used a mixture containing DER 332 (7.0 g), DDSA (8.0 g), DER 732 (3.2 g) and DMP-30 (0.3 g) for embedding amphibian eggs.

(ii) *Mixtures of epoxy resins.* In embedding media containing two epoxy resins the hardness of the block depends on the relative amounts of the two resins (Kushida 1959; Shinagawa et al. 1962).

TABLE 5.6

Variation of hardness with the ratio of two epoxy resins
(Shinagawa et al. 1962)

| Ingredients | (ml) | (ml) | (ml) | (ml) | (ml) |
|---|---|---|---|---|---|
| Epon 812 | 4.0 | 5.0 | 6.0 | 7.0 | 8.0 |
| Epon 815 | 6.0 | 5.0 | 4.0 | 3.0 | 2.0 |
| DDSA | 16.6 | 16.0 | 15.3 | 14.7 | 14.0 |
| DMP-30 | 0.53 | 0.52 | 0.51 | 0.49 | 0.48 |
| | soft ⟶ hard | | | | |

A ratio of Epon 812/Epon 815 of 5/5 is recommended for sectioning at room temperature. Softer blocks are suitable for sectioning in the cold, and harder blocks for sectioning in the warm.

(iii) *Hardeners.* The hardness of the block also depends on the nature of the hardener. The softest blocks result from the use of DDSA and harder blocks are obtained by replacing some or all of the DDSA with HHPA (Finck 1960) or MNA (Luft 1961; Robson 1964). Luft (1961) proposed using a mixture of Epon 812 and DDSA, and a mixture of Epon 812 and MNA in various proportions to obtain blocks of different hardness.

These mixtures are very popular since they provide an easy method for varying the hardness of the block.

The hardness also increases as the concentration of the accelerator increases, but adjustment of the accelerator is not a good method of varying the hardness since the block may become brittle and difficult to section.

TABLE 5.7

Variations of hardness with the ratio of DDSA/MNA (Luft 1961)

|  | (ml) | (ml) | (ml) | (ml) | (ml) |
|---|---|---|---|---|---|
| Mixture A | 10 | 7 | 5 | 3 | 0 |
| Mixture B | 0 | 3 | 5 | 7 | 10 |
| BDMA | 0.15 | 0.15 | 0.15 | 0.15 | 0.15 |
| | soft $\longrightarrow$ hard | | | | |

Mixture A: Epon 812 (62 ml) and DDSA (100 ml)
Mixture B: Epon 812 (100 ml) and MNA (89 ml)

(iv) *Epon-Araldite mixture.*    Mollenhauer (1964) recommended the three embedding media listed in Table 5.8, following a study of a range of different media consisting of various combinations of the epoxy resins Epon 812, Araldite CY 212 (M) and Araldite 506, the hardener DDSA, the flexibilizer Cardolite NC-513, the plasticizer dibutyl phthalate and the accelerators DMP-30 and BDMA.

TABLE 5.8

Composition and properties of embedding media containing Epon and Araldite (Mollenhauer 1964)

| *Mixture* | 1 | 2 | 3 |
|---|---|---|---|
| | (ml) | (ml) | (ml) |
| Epon 812 | 25 | 62 | – |
| Araldite CY 212 (M) | 15 | – | – |
| Araldite 506 | – | 81 | 50 |
| DDSA | 55 | 100 | – |
| Cardolite NC-513 | – | – | 25 |
| Dibutyl phthalate | 2–4 | 4–7 | 1–2 |
| *Relative hardness* | medium | soft–medium | soft–medium |
| *Image contrast* | high | medium | low |
| *Tissue preservation* | good | excellent | excellent |

These mixtures were developed primarily for plant tissues, and Mollenhauer (1964) suggested that changes in the proportions of the components of each mixture may be necessary to adapt them to a given tissue. In practice these mixtures have proved to be popular for a wide range of specimens with little modification, particularly mixtures Nos. 1 and 2.

If Araldite CY 212 is not available, Araldite 502 or Araldite 6005 can be substituted for it in the same proportions with equal success. Tissue embedded in mixture No. 1 is easier to section than when embedded in mixtures containing only Epon 812, Araldite CY 212, 502 or 6005. Tissue preservation is comparable to that obtained with Epon 812 alone. Mixture No. 2 is easier to section than No. 1 and gives better tissue preservation, but shows slightly less image contrast. Mixture No. 3 is slightly more difficult to section than either No. 1 or No. 2, and image contrast is lower. Preservation of tissue is excellent, and the mixture is particularly useful for specimens, such as pollen grains, which tend to be pulled out of the block during sectioning. This type of damage is rarely seen with mixture No. 3.

The plasticizer, dibutyl phthalate, improves the cutting properties of these mixtures, but can be eliminated if a relatively hard block is desired. Either DMP-30 or BDMA can be used as the accelerator for mixtures Nos. 1 and 2. BDMA must be used for mixture No. 3, since DMP-30 causes a precipitate to form slowly in the first epoxy resin/propylene oxide mixture during infiltration (see §5.3.4). The amount of accelerator required is about 1.5% of (fresh) DMP-30 and 3% of BDMA.

### 5.3.2c *Sectioning properties*

(i) *Anhydride/epoxy ratio.* The sectioning properties of the final block depend not only on its hardness but also on the extent and nature of the cross-links formed during polymerization. It is generally agreed that a linear, shorter polymer with few cross-links is easiest to section (Luft 1961; Coulter 1967), although there is some disagreement about the means of achieving this. Luft (1961) suggested that adjustment of the curing schedule will influence the sectioning properties, but Coulter (1967) found that this had no effect, and that the ratio of anhydride to epoxy was the important factor. The sectioning properties improved as the ratio was lowered from 1.0 to 0.6 and Coulter concluded that a reduction in anhydride favours a shorter polymer with fewer cross-links. The recommended modifications to Luft's (1961) Epon 812 embedding medium (Table 5.7) are as follows:

TABLE 5.9

Adjustment of the anhydride/epoxy ratio to
improve sectioning properties (Coulter 1967)

| | |
|---|---|
| Mixture A | Epon 812 (100 ml) and DDSA (124 ml) |
| Mixture B | Epon 812 (100 ml) and MNA (67 ml) |

The embedding medium consists of A and B mixed in the proportion of 1.5 to 1.0 with the addition of 1.5% DMP-30.

Using the same principles Coulter (1967) also calculated that an embedding medium based on Araldite 502 should contain 7.8 ml of DDSA to each 10 ml of Araldite 502 with 1.5% of DMP-30.

The calculations made by Coulter (1967) assumed that the weight per epoxide (WPE) (that is the number of grams of resin which contain an average of 1.0 g equivalent of epoxide) is 159 for Epon 812, and 502 for Araldite 502. Burke and Geiselman (1971) have found, however, that commercial samples of Epon 812 have considerable variations in WPE (from 140 to 160), and suggest that the correct proportions of hardener to epoxy resin should be calculated for each batch of resin. To be able to do this it is necessary to know the WPE; this is printed on the bottle by some suppliers (e.g. Ladd, see Appendix), and will be provided by other suppliers on request. Burke and Geiselman (1971) describe a method of calculating the correct proportions of Epon 812, DDSA and MNA to use with any particular batch of resin, and provide a table listing these proportions for any particular WPE. A typical mixture contains 27.1 g of Epon 812, 7.6 g of DDSA and 15.3 g of MNA, with the addition of 0.5 to 1.5% of BDMA or DMP-30. To take full advantage of these calculations it is essential to measure the weights of the various components very accurately. This is very difficult when the components are viscous and it is doubtful whether it is necessary for such measures to be taken in the majority of routine studies. It is suggested that electron microscopists only resort to such calculations when serious sectioning difficulties are encountered.

(ii) *Softening temperature*

TABLE 5.10

Addition of TAC to improve sectioning properties (Winborn 1965)

| Mixture | 1 | 2 | 3 |
|---|---|---|---|
| *Ingredient by volume* | (%) | (%) | (%) |
| DER 334 | 48 | – | – |
| Maraglas 655 | – | – | 48 |
| Araldite 502 | – | 68 | – |
| DDSA | 40 | 19 | - |
| Cardolite NC-513 | – | – | 40 |
| TAC | 10 | 10 | 10 |
| DMP-30 | 2 | 3 | 2 |
| *Relative hardness* | soft-medium | soft-medium | medium |
| *Image contrast* | high | high | high |
| *Thermal stability* | excellent | good | excellent |

Sectioning problems can also arise if the block softens at normal laboratory temperatures, and Winborn (1965) suggest that triallyl cyanurate (TAC) should be added to epoxy resin embedding media to give polymers with a high softening temperature. The nature of the reaction between TAC and the other components of the embedding medium is not known.

The hardness of the block decreases as the concentration of TAC increases. The mixture containing Maraglas 655 (No. 3) yields blocks which are slightly harder and more difficult to section than blocks containing DER 334 or Araldite 502.

### 5.3.3  PREPARATION OF EPOXY RESIN EMBEDDING MEDIA

Epoxy resins and their hardeners are supplied in cans or bottles and can be stored indefinitely at room temperature so long as they are kept *dry*. Storage in glass-stoppered bottles is usually preferred. The epoxy resin ERL 4206 and the epoxy flexibilizer DER 736 will dissolve the plastic liners of some screw caps and should be stored in bottles with glass or cork stoppers (Spurr 1969). Some samples of the accelerators are less stable and may be inactivated if they are allowed to take up water. They should be stored in bottles with well-fitting caps in a desiccator at room temperature.

**Note:**  Epoxy resins, hardeners and, in particular, accelerators are toxic and may cause skin irritation and dermatitis. They should be handled with great care in an area with good ventilation and preferably in a fume cupboard. Repeated contact with the skin should be avoided. The polymerized resin is inert.

Most of the epoxy resin embedding media take a long time to harden at room temperature, that is they have a long 'pot life', but they gradually become more viscous. Consequently, it is advisable to prepare the embedding medium just before use when the specimens are in the final stages of dehydration.

It is essential to mix the components of the embedding medium very thoroughly to obtain uniform polymerization. Inadequate mixing is one of the main causes of sectioning problems with epoxy resins. The simplest way to obtain good mixing is to warm the epoxy resin, the hardener *and* the containers to be used for measuring and mixing to 60 °C (Glauert and Glauert 1958). The hardener HHPA is solid at room temperature and must warmed to 55–60 °C before mixing with the epoxy resin. The warm components are easily poured from the containers in which they have been stored into a (warm) graduated cylinder, for measurement by volume, or into a (warm) beaker, on the pan of a balance, for measurement by weight. The mixture is immediately poured into a (warm) conical flask containing the required amount of accelerator. In general slight variations in the relative amounts of the various components are not important, except for the accelerator which should be measured accurately. A simple method is to calibrate a Pasteur pipette in 'drops of accelerator per ml' and to keep it for adding the accelerator. Alternatively a plastic syringe can be used. BDMA is much less viscous than DMP-30 and is much easier to handle. Care must be taken not to add too much accelerator or the block may be brittle and difficult to section.

The warm embedding medium in the conical flask is mixed by swirling and/or stirring with a warm glass rod. A uniform mixture is obtained in a few minutes by this method and no special stirring apparatus is required. A few bubbles may form during stirring but these will go if the mixture is allowed to stand at 60 °C for a short time. If the components of the embedding medium are not warmed then a much longer mixing time is required and it is advisable to use one of the stirrers specially designed for the purpose (e.g. Zacks 1963).

It is very difficult and tedious to wash glassware after it has been in contact with epoxy resins and it is convenient to use disposable polyethylene containers whenever possible since the resins do not stick to polyethylene. Plastic syringes are useful when it is necessary to measure small volumes of the components accurately. If glassware is used it is advisable to keep a separate set of glassware for the purpose, and to only attempt to clean the container used for measuring the amounts of the components (measuring cylinder or beaker). The mixing procedure is unaffected by a thin film of hardened resin on the surfaces of the other containers. Cleaning should be done in acetone or ethanol as soon as possible after use.

It is best to prepare the complete mixture just before use but if necessary it can be stored at 4 °C in a bottle with a well-fitting cap for several weeks, or for many months at − 20 °C. During infiltration of the specimens the embedding medium is kept at room temperature and warmed to 60 °C and stirred well before each change of solution.

### 5.3.4  EMBEDDING SCHEDULES FOR EPOXY RESINS

At the end of dehydration the specimens will be immersed in the intermediate solvent (usually propylene oxide) (see §4.2) in small vials (Fig. 3.2).

#### 5.3.4.a  *Standard embedding schedule*
1.  Remove the solvent with a pipette and flush it down a sink in a fume

cupboard with a large volume of water. If propylene oxide is being used do not remove it all since it is very volatile and the specimen may dry out.

2. Replace the solvent with a 1/1 mixture of solvent and embedding medium. Shake the vial gently to mix the components and leave for 30 min to 1 hr at room temperature.

3. Remove the mixture with a pipette and replace it with embedding medium. Leave 16 to 24 hr at room temperature with the cap off the vial to allow any remaining solvent to evaporate. Infiltration of dense specimens is aided by placing the vials on a shaker (Figs. 5.1 and 5.2).

4. Transfer each specimen to a *dry* capsule and fill the capsules with embedding medium. Insert labels (Fig. 5.5).

5. Place the capsules in an oven overnight at 60 °C (§ 5.2).

This schedule is given as a guide line for embedding. It is widely used for embedding in Epon 812 or Araldite CY 212, but it is clear from the published work (see references in Table 5.2) that the embedding schedule is not critical and that considerable variations can be introduced with equally successful results for the great majority of specimens. The original schedules recommended for the various epoxy resin embedding media are listed in Table 5.11.

Epoxy resins should not be poured down a sink after use since they may cause a blockage. Used embedding medium should be drained into a small box (an old plate box is suitable) and allowed to harden before being discarded.

Most epoxy resin embedding media harden overnight at 60 °C and this is a convenient temperature to use. Approximately 2 days are usually required for hardening at 50 °C, while the time can be reduced to 8 hr or less by raising the temperature to 70 °C or higher (e.g. Coulter 1967; Spurr 1969). The sectioning properties of most blocks improves with time and the capsules can safely be left in the oven for some days after hardening.

Luft (1961) suggested that a series of increasing temperatures (overnight at 35 °C, a day at 45 °C, and then overnight at 60 °C) be used for polymerizing epoxy resins, but there is little evidence that the resultant blocks are any different from those placed directly at 60 °C.

Ultra-violet (UV) irradiation can be used as an alternative to heat for the polymerization of epoxy resins (Shinagawa et al. 1962) with the advantage that polymerization can be carried out at a lower temperature if necessary (§ 5.2). Shinagawa et al. (1962) compared the hardening times of a mixture of Epon 812 and Epon 815 with heat and with UV and found that the times

TABLE 5.11

Embedding schedules for epoxy resins

| | p.o./Epoxy (2/1) | (1/1) | (1/2) | Embedding medium | Polymerization | | Reference |
|---|---|---|---|---|---|---|---|
| Standard schedule | – | 30–60 min | – | 16–24 hr | overnight | 60°C | Parker (1972) |
| Epok 533 | – | 60 min | – | 1–2 hr; 1–2 hr | 50 hr | 50°C | Kushida (1963b) |
| Maraglas 655 | – | 30 min | – | 12 hr (at 10°C) | 24–48 hr | 60°C | Spurlock et al. (1963) |
| Maraglas 655 | – | 45 min | – | 1 hr; 2–3 hr | 17 hr | 52°C | Erlandson (1964) |
| DER 332 | – | 1 hr (at 45°C) | – | 1 hr (at 37°C) | 3 × 24 hr | 37°C, 45°C, 60°C | Lockwood (1964) |
| Epon 812 + Araldite | 1 hr | – | 1 hr | 2–4 hr | overnight | 40°C–80°C | Mollenhauer (1964) |
| Media + TAC<br>Araldite 502<br>Maraglas 655<br>DER 334 | – | 30 min | (1/3)<br>30 min | 1 hr | 4 hr<br>+ 24 hr<br>+ 8 hr<br>+ 20–40 hr | 45°C<br>65–70°C<br>65–70°C<br>65–70°C | Winborn (1965) |
| ERL 4206 | – | 30 min | (1/3)<br>30 min | 3–4 hr; overnight (large specimens only) | 8–16 hr | 70°C | Spurr (1969) |

*Key*: p.o., propylene oxide; TAC, triallyl cyanurate

were 60 hr and 48 hr respectively at 50 °C, 170 hr and 70 hr at 40 °C, and infinity and 100 hr at 30 °C.

The standard embedding schedule is adequate for the majority of specimens, but sometimes the specimen is found to be softer than the surrounding embedding medium as a result of incomplete infiltration. This is more likely to happen with large, dense specimens and when thick cell walls are present. Adequate infiltration can be achieved using some or all of the following modifications:

(a) Make sure that the dehydration is complete by increasing the time of soaking in the intermediate solvent.

(b) Use an embedding medium of low viscosity (e.g. Spurr 1969).

(c) Pass the specimens more gradually from the intermediate solvent to the embedding medium using ratios of 2/1, 1/1, 1/2, etc.

(d) Allow a longer infiltration time at stage 3 of the standard schedule and place the vials on a shaker. Change the embedding medium at the beginning and end of each day.

(e) De-gas the specimens in a vacuum desiccator for at least 30 min before polymerization (Erlandson 1964) or reduce the pressure in the oven to 0.5 atmospheres for a part or all of the incubation period (Winborn 1965).

The final blocks can be stored indefinitely at room temperature and their sectioning properties continue to steadily improve. Bubbles tend to form between the gelatin capsule and hardened Maraglas when the capsules are left at room temperature for a long time. Spurlock et al. (1963) advise removing the blocks from the gelatin capsules soon after removal from the oven. Bubble formation does not occur with polyethylene capsules or with other epoxy resins, so long as the capsules are thoroughly dried before use.

### 5.3.4.b  *Embedding at room temperature*

The high temperatures used in the standard embedding schedule (§5.3.4a) are unsatisfactory for specimens for some histochemical studies. Kushida (1965c) developed a method of embedding at room temperature in a medium consisting of Epon 812 (100 ml), MNA (85 ml), Thiokol LP-8 (15–30 ml) and DMP-30 (3 ml) (Table 5.2). This mixture hardens in about 7 days at 20 °C, and in about 4 days at 30 °C. The temperature of the specimen is about 0.5 °C higher than that of the oven at 20 °C, and about 0.8 °C higher at 30 °C during the early stages of polymerization. Other epoxy resin embedding media can be polymerized at room temperature (approx. 22 °C) using ultra-violet irradiation (§5.3.4a).

Schedules for embedding at very low temperatures are described in (§6.1).

### 5.3.4c    *Rapid embedding*

The embedding procedure can be speeded up so long as the specimens are small (less than 0.1 mm in one dimension). Coulter (1967) used the rapid dehydration schedule described in §4.2.2 and then embedded specimens rapidly in Epon 812 or Araldite 502. The last change of propylene oxide was poured off, a 1:2 mixture of propylene oxide and epoxy resin added, and the specimens agitated for 5 min. The specimens were then transferred to the centres of capsules containing embedding medium which had been previously evacuated. The infiltration was completed in 5 min at the maximum vacuum available with a water aspirator. The embedding medium was polymerized at 95 °C for 40 min to 12 hr (for Araldite) or for 3 hr to 12 hr (for Epon).

Robbins and Jentzsch (1967) infiltrated monolayers of cells rapidly with Epon 812, using three 5 min changes of the complete embedding medium, following rapid dehydration in ethanol and Epon 812 (§4.2.3). Polymerization was not carried out at a high temperature and took 3 days at 55 °C.

Hayat and Giaquinta (1970) also developed a rapid embedding method with Epon 812 and found that adequate infiltration was obtained with two changes of Epon of 10 min each without the application of a vacuum. The embedding medium hardened in 1 hr at 100 °C and the blocks could be sectioned 15 min after removal from the oven.

## 5.4    Polyester resins

Polyester resins were introduced as embedding media for electron microscopy by Kellenberger et al. (1956) and Ryter and Kellenberger (1958a, b) in Switzerland and by Kushida (1960a) in Japan. They have similar excellent characteristics as embedding media to the epoxy resins but are less widely used because they are not so readily available and some of the components of the embedding medium are unstable and have to be replaced every few months.

### 5.4.1    POLYESTER RESIN EMBEDDING MEDIA

The earliest polyester embedding medium based on Vinox K3 (Kellenberger et al. 1956) was soon superceded by Vestopal W (Ryter and Kellenberger 1958a, b) which has remained the standard polyester embedding medium outside Japan. The embedding medium contains tertiary butyl perbenzoate or benzoyl peroxide (BP) as an initiator of polymerization and cobalt naphthenate (CN) as an accelerator (Table 5.12). BP and CN may explode if mixed together and this leads to problems in the preparation of the

TABLE 5.12

Polyester resin embedding media

| Polyester resin | Initiator or catalyst | | Activator or accelerator | | Comments | Reference |
|---|---|---|---|---|---|---|
| Vestopal W | TBP or BP | 1% | CN | 0.5% | First polyester resin embedding medium | Ryter and Kellenberger (1958a, b) |
| Vestopal W | TBP | 1% | CN | 0.5% | 10% TAC added to improve sectioning | Winborn (1963) |
| Vestopal W | BPP | 1% | – | | Activator omitted | Kushida (1964b) |
| Vestopal W | benzoin | 0.3% | – | | For polymerization by UV | Kushida (1964b) |
| Rigolac 2004 + Rigalac 70F | BPP | 1% | – | | Low viscosity. Ratio of 2004/70F determines hardness | Kushida (1960a) |
| Rigolac 2004 + Rigolac 70F | benzoin | 0.5% | – | | For polymerization by UV | Kushida (1961c) |
| Styrene/Rigolac 70F (7/3) or Styrene/Rigolac 2004 (6/4) | BP | 1% | – | | Ratio of styrene/Rigolac determines hardness | Shinagawa and Uchida (1961) |
| Rhodester 1108 | Butanox | 1% | – | | Rhodester; available in Paris; contains styrene | Argagnon and Enjalbert (1964) |
| Beetle 4116 22.5 ml + Beetle 4134 7.5 ml | Butanox M 50 | 0.3 ml | NL 49/ST or Q2 | 0.3 ml 0.3 ml | For rapid embedding | Rampley and Morris (1972) |

*Key:*

TBP  tertiary butyl perbenzoate
BP  benzoyl peroxide
BPP  benzoyl peroxide paste containing 50% benzoyl peroxide in tricresyl phosphate (Kushida 1960a). Manufactured by Riken Goseijushi in Japan. Trade name Luperco ATC.
benzoin  phenyl-benzoyl carbinol
Butanox M 50  A mixture of methyl ethyl ketone peroxides in a phthalate plasticizer
CN  cobalt naphthenate
NL 49/ST  a styrene-based solution containing 1% cobalt
Q2  a vanadium-based solution
TAC  triallyl cyanurate
UV  ultra-violet irradiation

embedding medium (§5.4.2). Kushida (1964b) has shown that Vestopal W can be polymerized by ultra-violet irradiation with the addition of benzoin as a catalyst, with the advantage that polymerization can be carried out at room temperature. Kushida (1964b) also obtained polymerization by heat with benzoyl peroxide in the absence of cobalt naphthenate. Winborn (1963) recommended the addition of 10% triallyl cyanurate to the Vestopal W embedding medium of Ryter and Kellenberger (1958a) to improve the sectioning properties of the final block.

The polyester resins Rigolac 2004 and Rigolas 70F manufactured in Japan (Kushida 1960a) also polymerize when heated in the presence of benzoyl peroxide alone, and can also be polymerized by UV irradiation (Kushida 1961c). The Rigolac mixture has the advantage that it is less viscous than Vestopal W and the hardness of the final block can be adjusted by varying the ratio of Rigolac 2004:Rigolac 70F from 8:2 to 6:4. The block has a yellow fluorescence which increases with the degree of polymerization and this is a convenient means of ascertaining how the polymerization is progressing (Kushida 1961c).

The other polyester resin embedding media described in the literature are Selectron (Low and Clevenger 1962) which was used in a mixture with n-butyl methacrylate (see §5.5), Rhodester (Argagnon and Enjalbert 1964), which is based on components locally available in France, and Beetle (Rampley and Morris 1972), which was recommended for rapid embedding since it hardens much faster than Vestopal W. The Beetle embedding medium was only introduced recently and it will be interesting to see if it becomes widely used.

A mixture of styrene and Rigolac (Table 5.12) can also be used as an embedding medium (Shinagawa and Uchida 1961). The hardness of the final block can be varied by altering the styrene/Rigolac ratio.

### 5.4.2  PREPARATION OF POLYESTER RESIN EMBEDDING MEDIA

Polyester resins, unlike epoxy resins, are polymerized by light, heat and oxygen (Kushida 1960b). They should be stored in bottles with tightly-fitting caps in a refrigerator and well protected from the light. They are then stable for many months. Benzoyl peroxide, cobalt naphthenate and the other catalysts and accelerators are easily inactivated and should be kept dry and away from the light in a refrigerator. Many workers recommend replacing stocks of the initiator and the accelerator every two months.

In general it is preferable to prepare the embedding medium just before use, although Kurtz (1961) reported that the complete Vestopal embedding mixture can be stored for a week in the refrigerator and that it will not polymerize at room temperature for 4 days.

As with epoxy resins (§5.3.3) thorough mixing of the viscous embedding medium is essential, and the use of a stirrer or mixer is advised (§5.2). When preparing an embedding medium containing cobalt naphthenate, the initiator (benzoyl peroxide) should be mixed well with the Vestopal W for 20 to 30 min before the addition of the cobalt naphthenate. The mixture is then stirred for a further 20 to 30 min (Estes and Apicella 1969). These precautions are necessary because benzoyl peroxide and cobalt naphthenate may explode if mixed directly together. Similarly the catalyst and accelerator in the Beetle embedding medium (Table 5.12) must not be mixed together before being added to the polyester resin (Rampley and Morris 1972).

### 5.4.3 EMBEDDING SCHEDULES FOR POLYESTER RESINS

Polyester resins are not soluble in ethanol and it is customary to dehydrate specimens in carefully dried acetone. Alternatively the specimens are dehydrated in ethanol and then acetone or styrene (Kurtz 1961; Shinagawa and Uchida 1961) is used as an intermediate solvent. Two changes of solvent of 30 min each are usual. Methyl methacrylate can also be used as an intermediate solvent (Kushida and Fujita 1968) with the advantage that it reacts with Vestopal W plus benzoyl peroxide and becomes an integral part of the polymer.

Polyester resins are viscous and are handled in the same way and with the same precautions as epoxy resins (§5.3.4). A shaker or agitator should be used to aid infiltration (§5.2) which is done at room temperature.

5.4.3.a  *Embedding schedule for Vestopal W*
(Ryter and Kellenberger 1958b; Kushida 1964b)

| | | |
|---|---|---|
| acetone/Vestopal W | 3/1 | 30–60 min |
| acetone/Vestopal W | 1/1 | 30–60 min |
| acetone/Vestopal W | 1/3 | 30–60 min |
| Vestopal W + catalyst | | several hours |
| polymerization | | 12–24 hr at 60°C (BP + CN as catalysts) |
| | *or* | 10 hr at room temperature under UV (benzoin as catalyst). |

For large or dense specimens better infiltration is obtained by a more gradual replacement of the acetone with Vestopal. The specimens are placed in a vial with a few drops of acetone and then the Vestopal is added drop by drop (De Haller et al. 1961). One drop of Vestopal (plus catalyst) is added every 30 min and is mixed well by shaking the vial. After 10 to 12 drops have been added, the Vestopal is replaced with fresh Vestopal and this is left overnight at room temperature. The next day the Vestopal is again replaced with fresh Vestopal and this is left for several hours before the specimens are embedded in capsules in the usual way.

Vestopal is more readily soluble in styrene than in acetone and when styrene is the intermediate solvent a soak in a 1/1 mixture of styrene/Vestopal W for 30 min under a moderate vacuum, followed by several hours (or overnight) in Vestopal W plus catalyst, is sufficient to obtain adequate infiltration (Kurtz 1961).

### 5.4.3.b    *Rapid embedding with Vestopal W*
(Estes and Apicella 1969)

Following dehydration in ethanol and two rinses in styrene of 5 min each:

| | | |
|---|---|---|
| Styrene/Vestopal W | 1/1 | 10 min |
| Vestopal W + BP + CN | | 10 min |
| Vestopal W + BP + CN | | 10 min |
| Polymerization | | 1 hr at 45–50 °C in a vacuum oven (15″ Hg) |
| | then | 15 min at 70–80 °C |
| | then | 15–20 min at 90–100 °C |

The blocks are ready for sectioning as soon as they have cooled to room temperature.

The Beetle embedding medium developed by Rampley and Morris (1972) is suitable for rapid embedding since the mixture polymerizes in 1 hr at 60 °C.

### 5.4.3.c    *Embedding schedule for Rigolac*
(Kushida 1960a, 1961c)

| | | |
|---|---|---|
| acetone/Rigolac + catalyst | 1/1 | 1–2 hr |
| Rigolac + catalyst | | 1–2 hr |
| Rigolac + catalyst | | 1–2 hr |
| polymerization | | 18–24 hr at 55 °C (BP as catalyst) |
| | *or* | 4–10 hr under UV (benzoin as catalyst) |

5.4.3.d   *Embedding schedule for styrene/Rigolac*

(Shinagawa and Uchida 1961)

| | | |
|---|---|---|
| acetone/styrene | 1/1 | 30 min |
| styrene | | 30 min |
| styrene/Rigolac + BP | 7/3 | 1–2 hr |
| | *or* 6/4 | |
| polymerization | | overnight at 50–60 °C |


## 5.5   Methacrylates

The methacrylates were widely used as embedding media in electron micro-scopy from 1949 when they were introduced by Newman et al. until the development of media based on epoxy and polyester resins in the late 1950s (§5.3 and §5.4). The methacrylates have major disadvantages as embedding media. In particular they polymerize with considerable shrinkage (up to 20%) and are unstable in the electron beam (Reid 1974). Consequently they are no longer used as routine embedding media, although the water-soluble metha-crylates (§5.6.2) have proved to be valuable for cytochemical studies. Only a brief description of the methods of embedding in methacrylates is included here. The reader is referred to earlier textbooks (Pease 1964; Glauert 1965) for fuller details.

### 5.5.1   METHACRYLATE EMBEDDING MEDIA

Methacrylate embedding media consist of a mixture of *n*-butyl and methyl methacrylates with benzoyl peroxide, Luperco or 2,2-azo-bis-iso-butyronitrile (AIB) as a catalyst for polymerization by heat, or with AIB, benzoin or uranyl nitrate (Kushida 1962c) for polymerization by ultra-violet irradiation (see review by Kushida 1965b). The hardness of the final block is adjusted by varying the proportion of *n*-butyl to methyl methacrylate, the hardness increasing with the proportion of methyl methacrylate.

Impurities in the embedding medium, or even the specimen itself, can initiate polymerization of methacrylates, so that polymerization may start in one region before others. The addition of a small quantity of an initiator, uranyl nitrate (0.01%) (Ward 1958), to the embedding medium aids more uniform polymerization.

The large shrinkage that occurs during polymerization of methacrylates may damage the specimen, and it is therefore customary to use partially polymerized methacrylate for the final infiltration and embedding. Part of the shrinkage will then have occurred before the specimen is placed in the embedding medium. Unfortunately, partially polymerized methacrylate is very viscous and the advantage of easy infiltration by low viscosity methacrylate is lost.

The stability of methacrylate sections under electron bombardment is improved by the addition of the cross-linking agent divinyl benzene to the embedding medium. The polymerized methacrylate has a three-dimensional structure similar to epoxy and polyester resins (Kushida 1961b). The shrinkage during polymerization is not reduced by the addition of divinyl benzene (Kushida 1962b), and the blocks tend to split vertically during polymerization (Watson and Aldridge 1961). Fortunately this split never passes through the specimen.

Greater stability in the electron beam is also obtained by using a mixture of *n*-butyl methacrylate and styrene as the embedding medium (Kushida 1961a). Copolymerization of methacrylate and styrene probably occurs, leading to a cross-linked structure similar to that obtained with divinyl benzene (Mohr and Cocking 1968). Methacrylate-styrene mixtures also shrink during polymerization. The hardness of the final block can be adjusted by varying the ratio of *n*-butyl methacrylate to styrene from 7:3 to 4:6 (Kushida 1961a). DeLamater et al. (1971) describe a method of embedding in styrene with the addition of only 5–10% of *n*-butyl methacrylate.

Methacrylates can also be copolymerized with polyester resins (Low and Clevenger 1962).

### 5.5.2  PREPARATION OF METHACRYLATE EMBEDDING MEDIA

The methacrylates are clear liquids of low viscosity. They are volatile and have an unpleasant smell; they should always be handled in a fume cupboard and great care should be taken not to inhale the fumes which can be harmful. They are supplied with an inhibitor, hydroquinone, to prevent polymerization during transport and storage. This inhibitor is removed by shaking the methacrylate with sodium hydroxide in a separating funnel (Pease 1964; Glauert 1965), followed by distillation under vacuum, if divinyl benzene or styrene is to be added to the embedding medium. Methyl methacrylate is distilled at 45 °C and 100 mmHg pressure, and *n*-butyl methacrylate at 52 °C and 11 mm Hg pressure (Kushida 1961b).

It is essential that the methacrylate be kept completely dry; it should be stored in bottles with tight-fitting lids and containing a drying agent (silica gel, molecular sieve or anhydrous calcium chloride) at room temperature in a fume cupboard. Divinyl benzene and styrene are polymerized by light, heat and oxygen (Kushida 1961a, b) and should be stored in a refrigerator well protected from the light. Any space in the containers should be filled with nitrogen. Styrene is supplied with an inhibitor, but this is not removed. More reproducible results are obtained with the inhibitor still present (Mohr and Cocking 1968).

Methacrylate embedding media mix easily because of their low viscosity. For a standard medium the *n*-butyl and methyl methacrylates are mixed in the proportions required (e.g. 8:2) to give the desired hardness to the final block, and then the catalyst (e.g. 1–2% benzoyl peroxide) is added and mixed well. Most catalysts contain water and it is important to dry the embedding medium before use. The complete embedding medium plus catalyst can be stored for some weeks at 4 °C.

Partially polymerized methacrylate is prepared by heating the methacrylate plus catalyst in a conical flask at 90 °C in a water bath until the mixture thickens. The mixture is agitated by continuously shaking the flask and the flask is immediately transferred to cold water when the mixture has the required viscosity.

Embedding media containing divinyl benzene are prepared with *n*-butyl and methyl methacrylate (100 ml), divinyl benzene (5 ml) and benzoyl peroxide (1 g) (Kushida 1961b), or with *n*-butyl methacrylate (distilled) (100 ml), divinyl benzene solution (3 ml), α-terpineol (3 ml) and benzoyl peroxide (1 g) (Watson and Aldridge 1961).

Methacrylate-styrene embedding media are prepared by mixing distilled *n*-butyl methacrylate and styrene in the required ratio (e.g. 7:3) and then adding 1% benzoyl peroxide as catalyst. The embedding medium is dried and filtered before use.

### 5.5.3 EMBEDDING SCHEDULES FOR METHACRYLATES

Methacrylates are readily soluble in ethanol and acetone and no intermediate solvent is required. The standard embedding schedule is:

| | |
|---|---|
| ethanol/methacrylate + catalyst | 1 hr |
| methacrylate + catalyst | 1 hr |
| methacrylate + catalyst | 1 hr |
| polymerization | overnight at 60 °C |
| *or* | 24–48 hr under UV |

Methacrylates are volatile and so the capsules are filled completely and the caps placed on.

As with other embedding media, polymerization by ultra-violet (UV) irradiation has the advantage that it can be carried out at room temperature or below (Shinagawa and Uchida 1961).

A similar infiltration schedule is used for embedding media containing divinyl benzene or styrene. With divinyl benzene polymerization is carried out for 24 hr at 50 °C (Kushida 1961b) or for 12 hr under UV at room temperature (Watson and Aldridge 1961). Methacrylate–styrene embedding media require 30 hr under UV irradiation (Kushida 1962a) or 18–24 hr at 55 °C for polymerization if the inhibitor has been removed from the styrene (Kushida 1961a) or 44 hr at 55 °C if the inhibitor is still present (Mohr and Cocking 1968). The slowing down of the polymerization by the inhibitor is an advantage as it permits thorough infiltration before hardening begins.

A longer infiltration procedure lasting 24 hr or more is required when using viscous, partially polymerized methacrylate. A higher temperature of polymerization (70–80 °C) is recommended to reduce the viscosity of the embedding medium and aid infiltration.

## 5.6    Water-soluble embedding media

Embedding media that are soluble in water can be used as dehydrating agents thus avoiding the use of ethanol, acetone and propylene oxide and so preserving some components of the specimen which would otherwise be extracted. However, these embedding media also cause extraction, although the components extracted may be different. In addition, the embedding media may react with the specimen and modify it. For example, epoxy compounds in aqueous solution react readily with proteins and nucleic acids and therefore tend to act as fixatives (Gibbons 1959). This makes water-soluble epoxy resins unsuitable for experiments on the digestion of thin sections by specific enzymes, and the water-soluble methacrylates are usually preferred (Bernhard 1966). However, these methacrylates are powerful lipid solvents for neutral lipids (Cope and Williams 1968) and for some phospholipids (Cope and Williams 1969b) even at low temperatures ($-20$ °C).

### 5.6.1    WATER-SOLUBLE EPOXY RESINS

The first water-soluble epoxy resin embedding medium to be developed was Aquon (Gibbons 1959), the water miscible fraction of Epon 812. Aquon is not available commercially and has to be extracted from Epon 812 in the laboratory by a rather lengthy procedure (see Glauert 1965 for the method of extraction). Consequently, it has been largely superceded by the Ciba epoxy resin Durcupan marketed by Fluka (see Appendix). The original

Water-soluble epoxy resin embedding media

| Epoxy resin | | Hardener | | Accelerator | | Additives | Comments | Reference |
|---|---|---|---|---|---|---|---|---|
| Aquon | 10 ml | DDSA | 25 ml | BDMA | 0.35 ml | – | | Gibbons (1959) |
| Aquon | 10 ml | DDSA + HHPA | 19 ml / 3 ml | BDMA | 0.30 ml | – | For harder specimens | Gibbons (1959) |
| Durcupan | 5 ml | DDSA | 11.7 ml | DMP-30 | 1.0–1.2 ml | DP 0.2–0.4 ml | DP not essential | Stäubli (1960); Leduc and Bernhard (1961) |
| Durcupan | | MNA | | DMP-30 | 1.5–2.0% | Cardolite NC-513 | Hardness varied with amounts of MNA and Cardolite | Kushida (1964a) |
| Durcupan | | DDSA + MNA | | DMP-30 | 1.5–2.0% | | Hardness varied with ratio of DDSA/MNA | Kushida (1964a) |
| Durcupan | 100 ml | MNA | 120 ml | DMP-30 | 1.5–2.0% | Thiokol LP-8 20–35 ml | Hardness varied with Thiokol | Kushida (1966a) |
| Epon 812 | 20 ml | HHPA | 16 ml | BDMA | 1.5–2.0% | | | Craig et al. (1962) |

Key:

Aquon A water-miscible fraction extracted from Epon 812.
Durcupan An aliphatic polyepoxide. A water-soluble Ciba epoxy resin originally named X133/2097.
DDSA dodecenyl succinic anhydride.
MNA methyl nadic anhydride
HHPA hexahydrophthalic anhydride
BDMA benzyl dimethyl amine
DMP-30 2,4,6-tridimethylaminomethyl phenol
DP dibutyl phthalate

For further details see the key to Table 5.2.

Durcupan embedding medium proposed by Stäubli (1960) (Table 5.13) is very difficult to section, and has been modified by Kushida (1964a, 1966a) to produce blocks with better sectioning properties. Kushida proposed the use of MNA as hardener with Cardolite NC-513 or Thiokol LP-8 as flexibilizer, or the use of a mixture of DDSA and MNA as hardener (Table 5.13). The hardness of the final block can be adjusted by altering the amounts of the various components of the embedding media.

A dehydration and embedding schedule based on Epon 812 (and not the water-soluble component Aquon) has been proposed by Craig et al. 1962.

The water-soluble epoxy resins are handled in exactly the same way and with the same precautions as other epoxy resins (§ 5.3.3 and § 5.3.4).

### 5.6.1a    *Dehydration and embedding schedule for Durcupan*
(Kushida 1964a)

A typical schedule is as follows:

| | | |
|---|---|---|
| 50% Durcupan in water | | 15–30 min |
| 70% Durcupan in water | | 15–30 min |
| 90% Durcupan in water | | 15–30 min |
| Durcupan | | 30–60 min |
| Durcupan | | 30–60 min |
| Durcupan/Durcupan + hardener + accelerator | 3/1 | 1 hr |
| Durcupan/Durcupan + hardener + accelerator | 1/1 | 1 hr |
| Durcupan/Durcupan + hardener + accelerator | 1/3 | 1 hr |
| Durcupan + hardener + accelerator | | 1–2 hr |
| Durcupan + hardener + accelerator | | 1–2 hr |
| Polymerization | | 50 hr at 50 °C |

The specimens are dehydrated and infiltrated with Durcupan at room temperature with constant agitation on a mechanical shaker (§ 5.2). 30% Durcupan in water should be avoided as the tissue tends to swell (Leduc and Bernhard 1961).

### 5.6.1b    *Dehydration in Durcupan followed by conventional embedding*
An alternative solution to the difficulty of sectioning the original Durcupan

embedding medium (Stäubli 1960) is to embed the specimens in a conventional embedding medium after dehydration in Durcupan. Embedding is still achieved without passing through highly reactive solvents such as propylene oxide. Specimens dehydrated in Durcupan have been successfully embedded in epoxy resins (Kushida 1963a; Stäubli 1963; Bird 1964) and polyester resins, methacrylate and styrene (Kushida 1965a). The specimens are passed from Durcupan to the embedding medium through a series of mixtures of increasing concentration of the embedding medium (e.g. 3/1, 1/1, 1/3 for 1 hr each). It is important to remove all the Durcupan when embedding in polyester resins, methacrylate or styrene, because it does not react with these embedding media and will soften the final block. Durcupan copolymerizes with other epoxy resins becoming part of the polymer. Following infiltration with the embedding medium polymerization is carried out in the usual way.

### 5.6.2 WATER-SOLUBLE METHACRYLATES

The first water-soluble methacrylate to be used as a combined dehydrating agent and embedding medium for electron microscopy was based on glycol methacrylate (GMA) (2-hydroxyethyl methacrylate) (Rosenberg et al. 1960; Leduc et al. 1963). In the period up to 1965 the quality of commercially available GMA varied considerably, and 2-hydroxypropyl methacrylate (HPMA) was investigated as an alternative (Leduc and Holt 1965). GMA of high quality is now available and is preferred to HPMA for most purposes because HPMA penetrates more slowly and the dehydration and embedding technique requires more care (Leduc and Bernhard 1967). GMA and HPMA are usually polymerized by irradiation with ultra-violet light (§ 5.2). A simple apparatus for this has been designed by Cole (1968). Methacrylates remain liquid at low temperatures and can therefore be used for dehydration and infiltration in the cold (§ 6.1).

### 5.6.2a *Glycol methacrylate*

Glycol methacrylate (GMA) is a clear, colourless liquid of low viscosity. It slowly polymerizes on storage and should be discarded if it is not readily miscible with water and a fresh lot obtained. It is not necessary to remove the inhibitor.

Blocks of pure GMA are very difficult to section and so *n*-butyl methacrylate (Leduc and Bernhard 1967) or styrene (Cope 1968) is added to the final embedding mixture to improve sectioning properties. Leduc and Bernhard (1967) recommended a mixture consisting of 7 parts of 97% GMA in water and 3 parts of *n*-butyl methacrylate (with inhibitor), containing 2%

Luperco (benzoyl peroxide paste) as catalyst, while Cope (1968) used GMA/ n-butyl methacrylate (7/3) or GMA/styrene (7/3) with 1.2% benzoyl peroxide as catalyst.

Partially polymerized embedding medium is used for the final infiltration and embedding and should have the viscosity of a thick syrup at 1–3 °C (Leduc and Bernhard 1967).

*Schedule for dehydration and embedding in GMA*
   (Cope and Williams 1968)

All procedures are carried out at 0–4 °C

| | |
|---|---|
| 80% GMA in water | 15 min |
| 100% GMA | 4 changes, 15 min each |
| embedding medium + catalyst | 1 hr |
| embedding medium + catalyst | 1 hr |
| polymerization in partially | |
| polymerized embedding medium | 48 hr under UV light at 20 °C |

Gelatin and not polyethylene capsules should be used.

Glycol methacrylate can also be used for dehydration before embedding in a polyester or epoxy resin (Kushida 1964b), and GMA can be copolymerized with Epon or Vestopal (McGee-Russell and De Bruijn 1964) to yield blocks with excellent sectioning properties.

### 5.6.2b  Hydroxypropyl methacrylate

Hydroxypropyl methacrylate (HPMA) is similar in appearance and properties to GMA; it should not be used unless it is completely miscible with water in the ratio of about 4:1 at 20 °C (Leduc and Holt 1965). The miscibility of water with HPMA *increases* as the temperature is lowered below 20 °C. HPMA is polymerized by light, even in the absence of a catalyst, and should be stored in the dark. The inhibitor need not be removed before use.

The embedding medium consists of 80% or 90% HPMA in water with 0.1% azonitrile as catalyst. Final infiltration and embedding is done in partially polymerized methacrylate (§ 5.5.2).

*Schedule for dehydration and embedding in HPMA*
   (Leduc and Holt 1965)

All procedures are carried out in the cold room using a mechanical shaker.

| | |
|---|---|
| 85% HPMA in $H_2O$ | 1 hr |
| 85% HPMA in $H_2O$ | 1 hr |

| 97% *or* 80% HPMA in $H_2O$ | 1 hr |
| 97% *or* 80% HPMA in $H_2O$ | 1 hr |
| partially polymerized | |
| embedding medium | 1 hr |
| polymerization | 12 hr-24 hr under UV at 10 °C |
| | *or* 2–3 days at 56 °C |

Pease (1966a, b) used HPMA as an embedding medium following dehydration in ethylene glycol (§4.3.1), while Brinkley et al. (1967) used HPMA as an intermediate solvent for monolayers of cells grown in Falcon plastic flasks to be embedded in Epon 812. These flasks dissolve in propylene oxide (§5.8.2).

### 5.6.2c *Embedding in the presence of water*

The first water-soluble methacrylate embedding medium proposed by Rosenberg et al. (1960) contained a small amount of water (about 3%), and Bernhard and his colleagues (e.g. Bartl and Bernhard 1966; Bernhard 1966) have shown that large proportions of water can be included in the final block. For example, Bartl and Bernhard (1966) embedded aldehyde-fixed tissue in a mixture of various water-soluble methacrylates (20%) with 0.28 M sucrose in water (80%). Such blocks are suitable for sectioning at low temperatures.

## 5.7 Other embedding media

The majority of the embedding media already described in this chapter are suitable for general use. A considerable number of other embedding media have been developed for electron microscopy, but none of them is yet sufficiently well established to warrant description in detail. The more promising of these are listed below. Most of them are water-soluble and can be polymerized in the presence of water to yield hydrophilic blocks suitable for cytochemical and immunochemical studies (see Lewis et al. 1974).

### 5.7.1 GELATIN

Gelatin has been tested as a water-soluble embedding medium in a number of laboratories (see Moretz et al. 1969b for references). A bacteriological grade of gelatin is used and the tissue is dehydrated by passing through gradually increasing concentrations of gelatin from 10% to 40% or 70% in water, the final solution containing 2% glycerin. The tissue in 40% or 70% gelatin is allowed to dry partially at 37 °C for some hours, and then small pieces of the gel are cut out and dried in a vacuum desiccator. There is a very large shrinkage on hardening and many authors have found sectioning to be difficult. Bernhard and Nancy (1964) embedded specimens in gelatin for sectioning at low temperatures (§6.4).

### 5.7.2   UREA-ALDEHYDE

Mixtures of urea and aldehydes can be polymerized in the presence of substantial quantities of water, providing an opportunity to preserve lipid structures in *situ* (Peterson and Pease 1970a). Casley-Smith (1967) embedded formaldehyde-fixed tissues in a commercial sample of urea-formaldehyde, with ammonium chloride as catalyst, while Peterson and Pease (1970b) used a mixture of 10 ml of 50% glutaraldehyde and 4 g of urea, with oxalic acid as catalyst. There are considerable difficulties in controlling the polymerization so as to obtain adequate infiltration of the embedding medium into the tissue, and the final blocks are hydrophilic and difficult to section.

### 5.7.3   PROTEIN-ALDEHYDE

Attempts have been made to cross-link other proteins, but there are considerable technical difficulties to overcome. Moretz et al. (1969b) tested agar-formaldehyde and found it to be less satisfactory than gelatin, while Robertson and Parsons (1970) reported that resorcinol-formaldehyde is not suitable for routine use.

Farrant and McLean (1969) used mixtures of serum and egg albumins with glutaraldehyde, but failed to obtain infiltration of the albumin into cells and organelles. The albumin was thus acting as an 'encapsulating' medium, (§3.6.4) rather than as an embedding medium. Nicholson (1971) embedded isolated chloroplasts in bovine serum albumin (BSA)-glutaraldehyde by a modification of the method of Farrant and McLean and obtained some promising results. He reported that there was some separation of the chloroplasts from the BSA matrix, suggesting incomplete infiltration of the BSA.

Kuhlmann and Viron (1972) describe a method of using albumin cross-linked with glutaraldehyde to 'embed' material which has not been dehydrated as a preliminary to the preparation of ultrathin frozen sections.

### 5.7.4   POLYAMPHOLYTES

McLean and Singer (1964) proposed the use of cross-linked polyampholytes as water-soluble embedding media for electron microscopy. Polyampholytes were prepared by the copolymerization of comparable molar quantities of two monomers of opposite sign to yield a block studded with more or less alternating negative and positive charges. The anionic monomer was methacrylic acid and the cationic monomer was dimethylaminoethyl methacrylate, used with a small amount of a cross-linking monomer tetramethylene dimethacrylate. Although this embedding medium appeared to be promising for immunochemical studies when it was first introduced, it has not been widely used and more attention is now being paid to the protein-aldehyde embedding media (§5.7.2 and §5.7.3).

## 5.8 Special embedding methods

### 5.8.1 FLAT EMBEDDING

In the standard embedding method (§5.2) gelatin or polyethylene capsules are used as embedding moulds. All ultramicrotomes (Reid 1974) are designed with chucks to hold the resultant cylindrical blocks. However, long specimens, such as fibres, flat specimens, such as monolayers of cells, and specimens which are to be sectioned in a particular orientation are more conveniently embedded in a rectangular block. Special 'vice-type' chucks are available on most ultramicrotomes to hold such blocks.

It is essential that the embedding medium can be easily separated from the mould and consequently moulds made out of polyethylene, aluminium foil or silicone rubber are used. Any shallow polyethylene containers are suitable, including mini-cube ice trays (Robertson et al. 1963), small disposable beakers cut off about 1 cm from the bottom (McCombs et al. 1968), small flat bottomed vials (EFFA mini-vials, see Appendix), and weighing trays (Fig. 5.6). McKinney and Walz (1969) recommend commercial polyethylene

Fig. 5.6. Polyethylene weighing trays are suitable as moulds for flat embedding. The embedding medium easily separates from the tray after polymerization. A label is incorporated in the block near each specimen.

containers supplied by Nalge (see Appendix). These are 7/8 in in diameter and $1\frac{1}{2}$–2 in deep, and have friction-fit covers, making them suitable for volatile embedding media such as the methacrylates. The height of the container is reduced to about 3/8 in by cutting with a razor blade. The hardened block is extruded from the mould by pushing with the thumbs, and the container is then reusable. If there is any difficulty in removing the block, the container is cut open.

When the exact orientation of the specimen in the block is important, rectangular moulds of the required dimensions can be constructed from polyethylene strips by the method described by Cheney and Ashhurst (1966).

Alternatively, moulds for flat embedding are made out of aluminium foil. A flat bottom to the mould is obtained by wrapping the foil around a metal block of the required dimensions and then pressing the block onto a piece of plate glass (Stetler 1972). Kushida (1969b) describes a technique in which a polyethylene container is lined with foil and covered with a sheet of cellophane for flat embedding with polyester resins. The foil protects the polyethylene which dissolves in the monomer resin.

Commercially available silicone rubber moulds and vinyl cups (Anderson and Doane 1967) are also suitable for flat embedding.

A number of specimens can be embedded in a single mould. A label (typed or indian ink) is incorporated in the block near each specimen (Fig. 5.6).

### 5.8.2  MONOLAYERS OF CELLS

Monolayers of cells are embedded in one of three ways:
  (i)   The layer of cells is removed from the substrate before embedding and the cells are embedded in capsules by the standard method (§ 5.2).
 (ii)   The layer of cells is removed from the substrate after embedding by separating the hardened block from the substrate.
(iii)   The layer of cells is grown on a substrate which can be sectioned and the cells are not separated from the substrate.

The third of these methods is preferable since the cells are easily damaged during the separation procedure. This damage is reduced if the cells are grown on glass coated with a special layer, or on various synthetic substrates, such as Falcon plastic Petri dishes (Table 5.14). Details of the preparation of these substrates can be found in the original references. The methods of cleaning glass before the deposition of a special layer, and the method of sterilizing the substrate before the cells are grown, are both important to ensure easy separation of the cells from the substrate at a later stage.

TABLE 5.14

Substrates to enable the separation of cell monolayers

| | |
|---|---|
| glass–carbon | Robbins and Gonatas (1964), and others |
| glass–collagen | Heyner (1963) |
| glass–collodion | Meiselman et al. (1967); Flaxman et al. (1968) |
| glass–collodion–collagen | Descarries (1967) |
| glass–formvar | Storb et al. (1966) |
| glass–formvar–carbon | Breton-Gorius (1968) |
| glass–gelatin | Speirs and Turner (1966) |
| glass–silica | Kushida and Suzuki (1968) |
| glass–silicone | Gay (1955); Rosen (1962) |
| glass–Teflon | Chang (1971) |
| glass–tungsten oxide | Kushida and Suzuki (1970) |
| mica | Persijn and Scherft (1965); Yardley and Brown (1965); Kumagewa et al. (1968) |
| Nunclon petri dish | Biberfield (1968) |
| polystyrene (Falcon Petri dish) | Goldberg et al. (1963); Brinkley et al. (1967); Zagury et al. (1968); Ross (1972) |
| polypropylene | Zagury et al. (1968) |
| polyester (Melinex) | Firket (1966) |
| polyester–carbon | Clegg (1964) |

5.8.2a *Removal of cells from the substrate before embedding*
When the orientation of the cell layer is not important, it can be removed from the substrate by scraping with a rubber 'policeman' or a razor blade before embedding. The under surfaces of the cells are inevitably damaged during removal, but this damage can be reduced if the cells are fixed *in situ* before removal and if they are grown on a coverslip coated with a layer of collodion or carbon (e.g. Daniel et al. 1966). The break then tends to occur in this layer. Some cell damage always occurs during mechanical removal and it is preferable to remove the cells more gently. When a collodion layer is present, the collodion and cells can be floated off onto buffer after fixation, (Meiselman et al. 1967). Alternatively the monolayer is dehydrated in the usual way and the collodion detaches from the glass in absolute ethanol (Descarries 1967) or propylene oxide (Flaxman et al. 1968), releasing the cell layer. Cells grown on plastic Petri dishes are also released in propylene oxide with the help of slight agitation (Biberfield 1968), since these dishes dissolve in propylene oxide.

5.8.2b *Separation of hardened blocks from the substrate; 'inverted capsule' technique*
Before the development of chucks on ultramicrotomes for holding flat blocks

it was customary to use the 'inverted capsule' technique to separate the hardened block containing the cells from the substrate. The monolayer of cells
on the substrate is dehydrated and infiltrated with embedding medium
in the same container in which it was fixed (§3.5). Propylene oxide is not
used when the cells are grown in plastic Petri dishes because the dishes dissolve. Hydroxypropyl methacrylate is a convenient alternative intermediate
solvent if one is required (Brinkley et al. 1967). The monolayer is then placed
face-upwards in a plastic Petri dish and capsules filled to the brim with embedding medium are placed over selected regions of the monolayer. Individual cells can be selected by viewing the monolayer before (Barnicot and
Huxley 1965; Lavail 1968) or after (Robbins and Gonatas 1964) fixation
with a light microscope. A circular mark is made on the back of the substrate
with a diamond-tipped slide marker or with glass marking ink (of a type not
removed in the subsequent dehydration).

Ordinary gelatin capsules can be used directly. BEEM polyethylene capsules are modified by removing the cap and cutting off the tip (Lavail 1968;
Branson 1971). The resultant cylinder is glued into position before being
filled with embedding medium.

The embedding medium is then polymerized by heat or ultra-violet (UV)
irradiation in the usual way (§5.2). When sheets of polyester film (Melinex)
are used as a substrate, the application of heat is avoided since it may cause
crystallization of the Melinex (Clegg 1964).

Various techniques have been tested for the separation of the hardened
blocks from the substrate. The capsules can be removed from some substrates
with a sharp pull after excess embedding medium has been trimmed away
from the base of the capsule with a razor blade. These substrates include
glass–silicone (Rosen 1962), glass–collagen (Heyner 1963), glass–carbon
(Robbins and Gonatas 1964; Barnicot and Huxley 1965; Lavail 1968),
polyester (Melinex) (Firket 1966), polystyrene (Falcon Petri dishes) (Zagury
et al. 1968) and polypropylene (Zagury et al. 1968).

It is easier to remove methacrylate blocks than blocks of epoxy resins,
and blocks of epoxy resins are more readily removed if this is done before the
embedding medium had hardened completely. For example, partially polymerized blocks of Epon (12–16 hr at 60–65 °C) are easily separated from
Falcon Petri dishes (Zagury et al. 1968) or glass (Branson 1971). The capsules
should be pulled off immediately after removal from the oven while they are
still warm. They are then returned to the oven to complete polymerization.

If it proves difficult to remove the capsules they can be loosened by cooling
the substrate by placing it on solid $CO_2$ (dry-ice) (Howatson and Almeida

1958). The capsules should immediately be placed in a desiccator to prevent water condensing on the surface and damaging the cells which are only covered with a very thin layer of embedding medium. Damage is often observed and cooling should only be used as a last resort.

When mica substrates are used the mica cleaves on shearing leaving a thin layer of mica attached to the end of the block (Persijn and Scherft 1965; Yardley and Brown 1965). This layer is removed during sectioning.

### 5.8.2c   *Separation of hardened blocks from the substrate; flat embedding technique*

When a vice-type chuck is available on the ultramicrotome for sectioning flat blocks, the monolayer on its substrate can be embedded in a flat embedding mould (§5.8.1). A thin layer of embedding medium is placed in the bottom of the mould and then the monolayer of cells on its substrate is placed face-downwards and covered with another thin layer of embedding medium. When the cells are grown in Falcon plastic Petri dishes, the dish, can be used as the embedding mould. Before polymerization most of the embedding medium is drained off leaving a thin layer covering the cells.

The resultant thin sheet containing the cells is separated from the substrate using one of the techniques used for capsules (§5.8.2b). The substrates that have proved suitable for removing sheets of cells without cooling include Falcon polystyrene dishes (Goldberg et al. 1963; Brinkley et al. 1967), polyester (Melinex)–carbon (Clegg 1964), glass–Teflon (Chang 1971) and glass–carbon (Langenberg et al. 1972).

Individual cells in monolayers growing on Melinex can be marked by inscribing a circle with a diamond scribe before fixation (Clegg 1964). The scribe indents the Melinex and the circle is visible on the surface of the hardened embedding medium after separation from the Melinex.

### 5.8.2d   *Cells not separated from the substrate*

When the substrate can be sectioned it does not have to be removed from the embedded layer of cells. Suitable substrates for this embedding method are listed in Table 5.15; they are ideal for studying the relationships between the cells and their substrate (Fig. 5.7). The substrate and cell layer are cut into strips and then embedded in capsules or they are embedded flat as described in §5.8.2c. Millipore membrane filters, silicone rubber membranes, Saran Wrap and Falcon Petri dishes soften or even dissolve in propylene oxide and another intermediate solvent is used. Millipore membrane filters are

TABLE 5.15

Substrates that can be sectioned

| | |
|---|---|
| Millipore membrane filters | Dalen and Nevalainen (1968) |
| | McCombs et al. (1968) |
| Nucleopore filters | Cornell (1969) |
| Saran Wrap | Holmes and Choppin (1968) |
| Silicone rubber membranes | Sharhar et al. (1972) |
| Methacrylate | Kjéllen et al. (1965) |
| Epon | Egeberg (1965) |
| | Zagury et al. (1966) |
| Araldite | Smith et al. (1969) |
| | Pegrum (1972) |
| Falcon plastic Petri dishes | Lipton and Konigsberg (1972) |
| | Nelson and Flaxman (1972) |

composed of mixed esters of cellulose and are available with pores of various dimensions (Friedman et al. 1968) (see Appendix). They can be used to fix and embed cells in suspension (McCombs et al. 1968) (§3.6.5), as well as monolayers. Friedman et al. (1968) and England (1969) studied the fine structures of Millipore filters. Nucleopore filters (Cornell 1969) have the advantage that they have flatter surfaces than Millipore filters with relatively few pores (Fig. 5.8).

Coverslips made out of embedding media (methacrylate, Epon or Araldite) make excellent substrates, but care has to be taken to remove all toxic factors

nf                                    0.5 μm

Fig. 5.7.   Section of a monolayer of chick cells growing on a nucleopore filter (nf). (Unpublished micrograph from a study by Audrey M. Glauert and Mary R. Daniel.)

Fig. 5.8. A chick cell has grown over a pore (p) in a nucleopore filter. (Unpublished micrograph from a study by Audrey M. Glauert and Mary R. Daniel.)

before the cells are grown. The simplest way to make the coverslips is to polymerize a very thin layer of embedding medium in a container with a flat, level bottom, such as a bowl of flexible polyvinyl chloride (Egeberg 1965), or a polyethylene ice-cube container (Smith et al. 1969) or an aluminium foil cooking tray (Parker 1972). After hardening the lower surface of the embedding medium can be polished with a metal polish to provide a coverslip with good optical properties for light microscopy (Pegrum 1972). These coverslips are ideal for the selection of individual cells for sectioning since they are suitable for viewing at the highest magnification of the light microscope.

In two recent studies (Nelson and Flaxman 1972; Lipton and Konigsberg

1972) it was shown that it is possible to section Falcon plastic Petri dishes. A rectangular or wedge-shaped block containing a slice of the dish and the overlying cells embedded in epoxy resin is cut out and sectioned directly.

### 5.8.3   EMBEDDING METHODS FOR VERY SMALL QUANTITIES OF MATERIAL

The simplest method of embedding very small quantities of material is to carry out the whole preparative procedure in a capsule as described in (§ 3.8). The main problem with this method is to obtain adequate infiltration of the embedding medium into the specimen. An embedding medium of low viscosity should be chosen and the capsules warmed in an oven ($50\,°C$–$60\,°C$) before the final stage of centrifugation to reduce the viscosity of the embedding medium (Blackburn 1968).

The lid of a BEEM polyethylene capsule is a suitable mould for the flat embedding of single cells (Bondareff and Hydén 1969).

REFERENCES

Anderson, N. and F. W. Doane (1967), Epoxy embedding of thin-layer material in commercially available vinyl cups for light and electron microscopy, Stain technol. *42*, 169.
Anderson, W. A. and J. André (1968), The extraction of some cell components with pronase and pepsin from thin sections of tissue embedded in an Epon-Araldite mixture, J. Microscopie *7*, 343.
Argagnon, J. and L. Enjalbert (1964), Technique d'inclusion pour la microscopie électronique utilisant un polyester: le Rhodester 1108 CPSL, J. Microscopie *3*, 339.
Barnicot, N. A. and H. E. Huxley (1965), Electron microscope observations on mitotic chromosomes, Q. Jl Microsc. Sci. *106*, 197.
Bartl, P. and W. Bernhard (1966), Essais d'inclusion de tissus dans des gels plastiques fortement hydratés, J. Microscopie *5*, 51.
Bernhard, W. (1966), Progress and limitations of ultrastructural cytochemistry carried out on ultrathin sections, J. Histochem. Cytochem. *14*, 746.
Bernhard, W. and M.-T. Nancy (1964), Coupes à congélation ultrafines de tissu inclus dans la gélatine, J. Microscopie, *3*, 579.
Biberfield, P. (1968), A method for the study of monolayer cultures with preserved orientation and interrelationship, J. Ultrastruct. Res. *25*, 158.
Bird, A. F. (1964), Embedding and staining small nematodes for electron microscopy, Nature, Lond. *203*, 1300.
Blackburn, W. R. (1968), Carrier centrifuge tube for gelatin and polyethylene embedding capsules, J. Ultrastruct. Res. *23*, 362.
Bluemink, J. G. (1970), The first cleavage of the amphibian egg, J. Ultrastruct. Res. *32*, 142.
Bondareff, W. and H. Hydén (1969), Isolation and embedding of single nerve cells for electron microscopy, Stain technol. *44*, 87.
Branson, S. H. (1971), Epon-embedded cell monolayers, Expl. Cell Res. *65*, 253.
Breton-Gorius, J. (1968), Étude des leucocytes étalés par coupes sériées parallèles et perpendiculaires au plan d'étalement, J. Microscopie *7*, 95.

Brinkley, B. R., P. Murphy and L. C. Richardson (1967), Procedure for embedding *in situ* selected cells cultured *in vitro*, J. Cell Biol. *35*, 279.

Burke, C. N. and C. W. Geiselman (1971), Exact anhydride epoxy percentages for electron microscopy embedding (Epon), J. Ultrastruct. Res. *36*, 119.

Casley-Smith, J. R. (1967), Some observations on the electron microscopy of lipids, Jl R. microsc. Soc. *87*, 463.

Chang, J. P. (1971), A new technique for separation of coverglass substrate from epoxy-embedded specimens for electron microscopy, J. Ultrastruct. Res. *37*, 370.

Cheney, R. A. and D. E. Ashhurst (1966), A method for the orientation of tissues in epoxy-resin blocks, and a design for a rotating stage for holding blocks during trimming, Jl R. microsc. Soc. *86*, 441.

Clegg, M. D. (1964), A technique for the microbeam irradiation of single cells in tissue culture, Jl R. microsc. Soc. *83*, 433.

Cole, M. B. (1968), A simple apparatus for ultraviolet polymerization of water-soluble embedding media employed in electron microscopy, J. Microscopie *7*, 441.

Cope, G. H. (1968), Low-temperature embedding in water-miscible methacrylates after treatment with antifreezes, Jl R. microsc. Soc. *88*, 235.

Cope, G. H. and M. A. Williams (1968), Quantitative studies on neutral lipid preservation in electron microscopy, Jl R. microsc. Soc. *88*, 259.

Cope, G. H. and M. A. Williams (1969a), Quantitative studies on the preservation of choline and ethanolamine phosphatides during tissue preparation for electron microscopy. I. Glutaraldehyde, osmium tetroxide, Araldite methods, J. Microscopy *90*, 31.

Cope, G. H. and M. A. Williams (1969b), Quantitative studies on the preservation of choline and ethanolamine phosphatides during tissue preparation for electron microscopy. II. Other preparative methods, J. Microscopy *90*, 47.

Cornell, R. (1969), The use of nucleopore filters in ultrastructural studies of cell cultures, Expl Cell Res. *56*, 156.

Coulter, H. D. (1967), Rapid and improved methods for embedding biological tissues in Epon 812 and Araldite 502, J. Ultrastruct. Res. *20*, 346.

Craig, E. L., W. J. Frajola and M. H. Greider (1962), An embedding technique for electron microscopy using Epon 812, J. Cell Biol. *12*, 190.

Dalen, H. and T. J. Nevalainen (1968), Direct epoxy embedding for vertical sectioning of cells grown as a monolayer on Millipore filters, Stain technol. *43*, 217.

Daniel, M. R., J. T. Dingle, A. M. Glauert and J. A. Lucy (1966), The action of excess of vitamin A alcohol on the fine structure of rat dermal fibroblasts, J. Cell Biol. *30*, 465.

De Haller, G., C. F. Ehret and R. Naef (1961), Technique d'inclusion et d'ultramicrotomie, destinée à l'étude du développement des organelles dans une cellule isolée, Experientia *17*, 524.

DeLamater, E. D., E. Johnson, T. Schoen and C. Whitaker (1971), The use of styrenes as embedding media for electron microscopy, Proc. 29th Ann. Conf. EMSA, p. 488.

Descarries, L. (1967), An improved method for preparing cultivated nervous tissue for electron microscopic study, J. Microscopie *6*, 313.

Egeberg, J. (1965), Sandwich embedding of monolayers of cells in epoxy resin for ultrathin sectioning, Stain technol. *40*, 343.

England, M. A. (1969), Millipore filters studied in isolation and *in vitro* by transmission electron microscopy and stereoscanning electron microscopy, Expl Cell Res. *54*, 222.

Erlandson, R. A. (1964), A new Maraglas, D.E.R. 732, embedment for electron microscopy, J. Cell Biol. *22*, 704.

Estes, L. W. and J. V. Apicella (1969), A rapid embedding technique for electron microscopy, Lab. Invest. *20*, 159.

Farrant, J. L. and J. D. McLean (1969), Albumins as embedding media for electron microscopy, Proc. 27th Ann. Conf. EMSA, p. 422.

Finck, H. (1960), Epoxy resins in electron microscopy, J. biophys. biochem. Cytol. 7, 27.

Firket, H. (1966), Polyester sheeting (Melinex O), a tissue culture support easily separable from epoxy resins after flat-face embedding, Stain technol. 41, 189.

Flaxman, B. A., M. A. Lutzner and E. J. Van Scott (1968), Ultrastructure of cell attachment to substratum *in vitro*, J. Cell Biol. 36, 406.

Freeman, J. A. and B. O. Spurlock (1962), A new epoxy embedment for electron microscopy, J. Cell Biol. 13, 437.

Friedman, B., P. Blais and P. Shaffer (1968), Fine structure of millipore filters, J. Cell Biol. 39, 208.

Gay, H. (1955), Serial sections of smears for electron microscopy, Stain technol. 30, 239.

Gibbons, I. R. (1959), An embedding resin miscible with water for electron microscopy, Nature, Lond. 184, 375.

Glauert, A. M. (1965), The fixation and embedding of biological specimens, in: Techniques for electron microscopy, D. H. Kay, ed. (Blackwell Scientific Publications, Oxford), p. 166.

Glauert, A. M. and R. H. Glauert (1958), Araldite as an embedding medium for electron microscopy, J. biophys. biochem. Cytol. 4, 191.

Glauert, A. M., G. E. Rogers and R. H. Glauert (1956), A new embedding medium for electron microscopy, Nature, Lond. 178, 803.

Goldberg, B., H. Green and G. J. Todaro (1963), Collagen formation *in vitro* by established mammalian cell lines, Expl Cell Res. 31, 444.

Hayat, M. A. and R. Giaquinta (1970), Rapid fixation and embedding for electron microscopy, Tissue and Cell 2, 191.

Heyner, S. (1963), *In situ* embedding of cultured cells or tissue, grown on glass, in epoxy resins for electron microscopy, Stain technol. 38, 335.

Holmes, K. V. and P. W. Choppin (1968), On the role of microtubules in movement and alignment of nuclei in virus-induced syncytia, J. Cell Biol. 39, 526.

Howatson, A. F. and J. D. Almeida (1958), A method for the study of cultured cells by thin sectioning and electron microscopy, J. biophys. biochem. Cytol. 4, 115.

Jurand, A. and M. J. Ireland (1965), A slow rotary shaker for embedding in viscous media, Stain technol. 40, 233.

Kellenberger, E., W. Schwab and A. Ryter (1956), L'utilisation d'un copolymère des polyesters comme matériel d'inclusion en ultramicrotomie, Experientia 12, 421.

Kjellén, L., G. Lagermalm, A. Svedmyr and K. G. Thorsson (1965), Crystalline-like patterns in nuclei of cells infected with an animal virus, Nature, Lond. 175, 505.

Korn, E. D. and R. A. Weisman (1966), Loss of lipids during preparation of amoebae for electron microscopy, Biochim. biophys. Acta 116, 309.

Kuhlmann, W. D. and A. Viron (1972), Cross-linked albumin as supporting matrix in ultrathin cryo microtomy, J. Ultrastruct. Res. 41, 385.

Kumagewa, M., M. Cattoni and G. C. Rose (1968), Electron microscopy of oral cells *in vitro*. III. *In situ* embedding of cultures in chambers of the circumfusion system, Tex. Rep. Biol. Med. 26, 205.

Kurtz, S. M. (1961), A new method for embedding tissues in Vestopal W, J. Ultrastruct. Res. 5, 468.

Kushida, H. (1959), On an epoxy resin embedding method for ultrathin sectioning, J. Electron Microscopy 8, 72.

Kushida, H. (1960a), A new polyester embedding method for ultrathin sectioning, J. Electron Microscopy 9, 113.

Kushida, H. (1960b), On the handling of epoxy resins and polyester resins for electron microscopy, J. Electron Microscopy 9, 157.

Kushida, H. (1961a), A styrene-methacrylate resin embedding method for ultrathin sectioning, J. Electron Microscopy 10, 16.

Kushida, H. (1961b), A new embedding method for ultrathin sectioning using a methacrylate resin with three dimensional polymer structure, J. Electron Microscopy *10*, 194.

Kushida, H. (1961c), On ultraviolet polymerization of polyester resins in embedding for electron microscopy, J. Electron Microscopy *10*, 201.

Kushida, H. (1962a), On ultraviolet polymerization of styrene resins in embedding for electron microscopy, J. Electron Microscopy *11*, 128.

Kushida, H. (1962b), A study of cellular swelling and shrinkage during fixation, dehydration and embedding in various standard media, J. Electron Microscopy *11*, 135.

Kushida, H. (1962c), Uranyl nitrate as a catalyst for ultraviolet polymerization in embedding, J. Electron Microscopy *11*, 253.

Kushida, H. (1963a), A modification of the water-miscible epoxy resin 'Durcupan' embedding method for ultrathin sectioning, J. Electron Microscopy *12*, 71.

Kushida, H. (1963b), An improved epoxy resin 'Epok 533', and polyethylene glycol 200 as a dehydrating agent, J. Electron Microscopy *12*, 167.

Kushida, H. (1964a), Improved methods for embedding with Durcupan, J. Electron Microscopy *13*, 139.

Kushida, H. (1964b), Glycol methacrylate as a dehydrating agent for embedding with polyester and epoxy resins, J. Electron Microscopy *13*, 200.

Kushida, H. (1965a), Durcupan as a dehydrating agent for embedding with polyester, styrene and methacrylate resins, J. Electron Microscopy *14*, 52.

Kushida, H. (1965b), Dehydration and embedding for electron microscopy. 2. Embedding, J. Electron Microscopy *14*, 251.

Kushida, H. (1965c), A new method for embedding with epoxy resin at room temperature, J. Electron Microscopy *14*, 275.

Kushida, H. (1966a), Further improved method for embedding with Durcupan, J. Electron Microscopy *15*, 95.

Kushida, H. (1966b), New embedding with D.E.R. 732 and Epon 812, J. Electron Microscopy *15*, 96.

Kushida, H. (1967), A new embedding method employing D.E.R. 736 and Epon 812, J. Electron Microscopy *16*, 278.

Kushida, H. (1969a), A new rotary shaker for fixation, dehydration and embedding, J. Electron Microscopy *18*, 137.

Kushida, H. (1969b), Flat embedding with polyester resins, J. Electron Microscopy *18*, 197.

Kushida, H. (1971), A new method for embedding with Epon 812, J. Electron Microscopy *20*, 206.

Kushida, H. and K. Fujita (1968), Methyl methacrylate as an auxiliary to infiltration for embedding with Vestopal W, J. Electron Microscopy *17*, 349.

Kushida, H. and K. Fujita (1971), New device for penetration of embedding media into specimens, J. Electron Microscopy *20*, 208.

Kushida, H. and K. Suzuki (1968), Evaporated silica as a stripping agent for coverslip-cultured cells embedded in epoxy resins, J. Electron Microscopy *17*, 350.

Kushida, H. and K. Suzuki (1970), Evaporated tungsten oxide as a stripping agent for monolayer-cultured cells embedded in epoxy resins, J. Electron Microscopy *19*, 191.

Langenberg, W. G., H. F. Schroeder, A. B. Welch and G. E. Cook (1972), Epoxy film separable from glass surfaces for selective light and electron microscopy of tissue and *in situ* grown cells, Stain technol. *47*, 303.

Lavail, M. M. (1968), A method of embedding selected areas of tissue cultures for electron microscopy, Tex. Rep. Biol. Med. *26*, 215.

Leduc, E. H. and W. Bernhard (1961), Ultrastructural cytochemistry. Enzyme and acid hydrolysis of nucleic acids and proteins, J. biophys. biochem. Cytol. *10*, 437.

Leduc, E. H. and W. Bernhard (1967), Recent modifications of the glycol methacrylate embedding procedure, J. Ultrastruct. Res. *19*, 196.

Leduc, E. H. and S. J. Holt (1965), Hydroxypropyl methacrylate, a new-water-miscible embedding medium for electron microscopy, J. Cell Biol. *26*, 137.

Leduc, E. H., V. Marinozzi and W. Bernhard (1963), The use of water-soluble glycol methacrylate in ultrastructural cytochemistry, Jl R. microsc. Soc. *81*, 119.

Lewis, P. R., D. P. Knight and M. A. Williams (1974), Staining methods for thin sections, in: Practical methods in electron microscopy, A. M. Glauert, ed. (North-Holland, Amsterdam).

Lipton, B. H. and I. R. Konigsberg (1972), A fine-structural analysis of the fusion of myogenic cells, J. Cell Biol. *53*, 348.

Lockwood, W. R. (1964), A reliable and easily sectioned epoxy resin embedding medium, Anat. Rec. *150*, 129.

Low, F. N. and M. R. Clevenger (1962), Polyester-methacrylate embedments for electron microscopy, J. Cell Biol. *12*, 615.

Luft, J. H. (1961), Improvements in epoxy resin embedding methods, J. biophys. biochem. Cytol. *9*, 409.

Maaløe, O. and A. Birch-Andersen (1956), On the organization of the 'nuclear material' in *Salmonella typhimurium*, Symp. Soc. gen. Microbiol. *6*, 261.

McCombs, R. M., M. Benyesh-Melnick and J. P. Brunschwig (1968), The use of Millipore filters in ultrastructural studies of cell cultures and viruses, J. Cell Biol. *36*, 231.

McGee-Russell, S. M. and W. C. De Bruijn (1964), Experiments on embedding media for electron microscopy, Q. Jl Microsc. Sci. *105*, 231.

McKinney, R. V. and E. J. Walz (1969), Flat embedding of thin tissues in commercially available polyethylene containers, Stain technol. *44*, 251.

McLean, J. D. and S. J. Singer (1964), Cross-linked polyampholytes. New water-soluble embedding media for electron microscopy, J. Cell Biol. *20*, 518.

Meiselman, N., A. Kohn and D. Danon (1967), Electron microscopic study of penetration of Newcastle disease virus into cells leading to formation of polykaryocytes, J. Cell Sci. *2*, 71.

Mohr, W. P. and E. C. Cocking (1968), A method of preparing highly vacuolated, senescent, or damaged plant tissue for ultrastructural study, J. Ultrastruct. Res. *21*, 171.

Mollenhauer, H. H. (1959), Permanganate fixation of plant cells, J. biophys. biochem. *6*, 431.

Mollenhauer, H. H. (1964), Plastic embedding mixtures for use in electron microscopy, Stain technol. *39*, 111.

Moretz, R. C., C. K. Akers and D. F. Parsons (1969a), Use of small angle X-ray diffraction to investigate disordering of membranes during preparation for electron microscopy. I. Osmium tetroxide and potassium permanganate, Biochim. biophys. Acta *193*, 1.

Moretz, R. C., C. K. Akers and D. F. Parsons (1969b), Use of small angle X-ray diffraction to investigate disordering of membranes during preparation for electron microscopy. II. Aldehydes, Biochim. biophys. Acta *193*, 12.

Nelson, B. K. and B. A. Flaxman (1972), *In situ* embedding and vertical sectioning for electron microscopy of tissue cultures grown on plastic Petri dishes, Stain technol. *47*, 261.

Newman, S. B., E. Borysko and M. Swerdlow (1949), New sectioning techniques for light and electron microscopy, Science *110*, 66.

Nicolson, G. L. (1971), Structure of the photosynthetic apparatus in protein-embedded chloroplasts, J. Cell Biol. *50*, 258.

Parker, R. A. (1972), Personal communication, Strangeways Research Laboratory, Cambridge.

Parsons, D. F. and E. B. Darden (1961), Optimal conditions for methacrylate embedding of certain tissues and cells sensitive to polymerization damage, Expl Cell Res. *24*, 466.

Pease, D. C. (1964), Histological techniques for electron microscopy, 2nd edition (Academic Press, New York and London).

Pease, D. C. (1966a), The preservation of unfixed cytological detail by dehydration with 'inert' agents, J. Ultrastruct. Res. *14*, 356.

Pease, D. C. (1966b), Anhydrous ultrathin sectioning and staining for electron microscopy, J. Ultrastruct. Res. *14*, 379.

Pegrum, S. M. (1972), Personal communication, University College, London.

Persijn, J. P. and J. P. Scherft (1965), Sheet mica – a nonadherent carrier for surface culture of cells to be embedded in Epon, Stain technol. *40*, 89.

Peterson, R. G. and D. C. Pease (1970a), Polymerizable glutaraldehyde-urea mixtures as water-soluble embedding media, Proc. 28th Ann. Conf. EMSA. p. 334.

Peterson, R. G. and D. C. Pease (1970b), Features of the fine structure of myelin embedded in water-containing aldehyde resins, Proc. 7th Int. Congr. Electron Microscopy, Grenoble, *1*, 409.

Rampley, R. D. and A. Morris (1972), A rapid method for polyester embedding, Proc. 5th. Eur. Reg. Conf. Electron Microscopy, Manchester, p. 224.

Reedy, M. K. (1965), Section staining for electron microscopy. Incompatibility of methyl nadic anhydride with permanganates, J. Cell Biol. *26*, 309.

Reid, N. (1974), Ultramicrotomy, in: Practical methods in electron microscopy, A. M. Glauert, ed. (North-Holland, Amsterdam).

Richardson, K. C., L. Jarett and E. H. Finke (1960), Embedding in epoxy resins for ultrathin sectioning in electron microscopy, Stain technol. *35*, 313.

Robbins, E. and N. K. Gonatas (1964), *In vitro* selection of the mitotic cell for subsequent electron microscopy, J. Cell Biol. *20*, 356.

Robbins, E. and G. Jentzsch (1967), Rapid embedding of cell culture monolayers and suspensions for electron microscopy, J. Histochem. Cytochem. *15*, 181.

Robertson, J. D., T. S. Bodenheimer and D. E. Stage (1963), The ultrastructure of Maunther cell synapses and nodes in goldfish brain, J. Cell Biol. *19*, 159.

Robertson, J. G. and D. F. Parsons, (1970), Myelin structure and retention of cholesterol in frog sciatic nerve embedded in a resorcinol-formaldehyde resin, Biochim. biophys. Acta *219*, 379.

Robson, E. A. (1964), The cuticle of *Peripatopsis moseleyi*, Q. Jl Microsc. Sci. *105*, 281.

Rosen, S. I. (1962), Cover-glass embedding in open-end capsules for electron microscopy, Stain technol. *37*, 195.

Rosenberg, M., P. Bartl and J. Leško (1960), Water-soluble methacrylate as an embedding medium for the preparation of ultrathin sections, J. Ultrastruct. Res. *4*, 298.

Ross, R. (1972), Personal communication, University of Washington, Seattle.

Ryter, A. and E. Kellenberger (1958a), Etude au microscope électronique de plasmas contenant de l'acide désoxyribonucléique, Z. Naturf. *13*, 597.

Ryter, A. and E. Kellenberger (1958b), L'inclusion au polyester pour l'ultramicrotomie, J. Ultrastruct. Res. *2*, 200.

Sharhar, A., O. Novick and Y. Straussman (1972), Utilization of silicon rubber membranes as supporting medium for cells for electron microscopy, Proc. 5th Eur. Reg. Conf. Electron Microscopy, Manchester, p. 226.

Shinagawa, Y. and Y. Uchida (1961), On the specimen damage of spinal cord due to polymerization of embedding media, J. Electron Microscopy *10*, 86.

Shinagawa, Y., S. Yahara and Y. Uchida (1962), Polymerization of epoxy resin for electron microscopy in the cold, J. Electron Microscopy *11*, 133.

Smith, W., E. W. Gray and J. M. K. Mackay (1969), A sandwich-embedding technique for monolayers of cells cultured on Araldite, J. Microscopy *89*, 359.

Speirs, R. S. and M. X. Turner (1966), Serial sampling of cell cultures using gelatin coated slides in coplin jars, Expl Cell Res. *44*, 661.

Spurlock, B. O., V. C. Kattine and J. A. Freeman (1963), Technical modifications in Maraglas embedding, J. Cell Biol. *17*, 203.

Spurr, A. R. (1969), A low-viscosity epoxy resin embedding medium for electron microscopy, J. Ultrastruct. Res. *26*, 31.

Stäubli, W. (1960), Nouvelle matière d'inclusion hydrosoluble pour la cytologie électronique, C. r. Séanc. Soc. Biol. *250*, 1137.

Stäubli, W. (1963), A new embedding technique for electron microscopy, combining a water-soluble epoxy resin (Durcupan) with water-insoluble Araldite, J. Cell Biol. *16*, 197.

Stein, O. and Y. Stein (1971), Light and electron microscopic radioautography of lipids: techniques and biological applications, Adv. Lipid Res. *9*, 1.

Steinbrecht, R. A. and K. D. Ernst (1967), Continuous penetration of delicate tissue specimens with embedding resin, Science Tools *14*, 24.

Stetler, D. A. (1972), Thin-layer embedding of filamentous plant parts for light and electron microscopy, Stain technol. *47*, 270.

Storb, R., R. L. Amy, R. K. Wertz, B. Fauconnier and M. Bessis (1966), An electron microscope study of vitally stained single cells irradiated with a ruby laser microbeam, J. Cell Biol. 31, 11.

Szubinska, B. (1971), 'New membrane' formation in *Amoeba proteus* upon injury of individual cells, J. Cell Biol. *49*, 747.

Voelz, H. and M. Dworkin (1962), Fine structure of *Myxococcus xanthus* during morphogenesis, J. Bact. *84*, 943.

Ward, R. T. (1958), Prevention of polymerization damage in methacrylate embedding, J. Histochem. Cytochem. 6, 398.

Watson, M. L. and W. G. Aldridge (1961), Methods for the use of indium as an electron stain for nucleic acids, J. biophys. biochem. Cytol. *11*, 257.

Winborn, W. B. (1963), Light and electron microscopy of the islets of Langerhans of the Saimiri monkey pancreas, Anat. Rec. *147*, 65.

Winborn, W. B. (1965), Dow epoxy resin with triallyl cyanurate, and similarly modified Araldite and Maraglas mixtures, as embedding media for electron microscopy, Stain technol. *40*, 227.

Yardley, J. and G. Brown (1965), Fibroblasts in tissue culture. Use of colloidal iron for ultrastructure localization of acid mucopolysaccharides. Lab. Invest. *14*, 501.

Zacks, S. I. (1963), Mechanical stirrer for epoxy resin embedding media, Stain technol. *38*, 60.

Zagury, D., G. D. Pappas and P. I. Marcus (1968), Preparation of cell monolayers for combined light and electron microscopy: staining in blocks, J. Microscopie *7*, 287.

Zagury, D., P. Zeitoun and M. Vielte (1966), Cytologie: méthode d'étude de cultures des cellules par la microscopie électronique, C. r. Acad. Sci. *262*, 1458.

# Chapter 6

# Low-temperature methods

Specimens are prepared at low temperatures with the aim of reducing the extraction of labile components by the fixative, dehydrating agent and embedding medium. In the simplest method a standard schedule is used but all the solutions are kept cold in an ordinary deep-freeze cabinet. The other low-temperature methods, freeze-drying, freeze-substitution and cryo-ultramicrotomy, are not sufficiently widely used to warrant detailed description here. They all involve an initial rapid freezing of the specimen and this is often the only means of fixation. Useful descriptions of the mechanism of freezing in biological systems are given by Bullivant (1970) and Rebhun (1972).

## 6.1 Fixation, dehydration and embedding at low temperatures

Fernández-Morán (1961) was one of the first to propose a low-temperature schedule for electron microscopy. Pieces of tissue were fixed at 0 °C, dehydrated in a graded series of acetone or methanol with the temperature progressively decreasing from 0 °C to − 25 °C, and then infiltrated with methacrylate at − 25 °C over a period of 3 to 5 days. The methacrylate was polymerized by ultra-violet (UV) irradiation (§5.2) at − 25 °C or at − 55 °C, the specimen temperature being below − 10 °C or − 40 °C respectively.

More recently Cope (1968) developed a method of low-temperature dehydration and embedding in water-miscible methacrylates (§5.6.2). The tissues were treated with a cryo-protective agent (dimethyl sulphoxide or dimethyl sulphoxide mixed with glycerol or ethylene glycol) and cooled to − 20 °C before fixation.

The following schedule was recommended:

(1) Cut up tissue in 10% antifreeze solution at 0 °C for 5 min.

(2) Transfer tissue to 25% antifreeze solution at 0 °C for 5 min.

(3) Transfer tissue to 50% antifreeze solution at 0 °C for 5 min.

(4) Cool tissue in 50% antifreeze solution slowly to − 20 °C by placing in a deep-freeze cabinet for 40 min.

(5) Fix tissue in 5% glutaraldehyde in 50% antifreeze solution at − 20 °C for 6 to 8 hr.

(6) Dehydrate tissue in 4 to 6 changes of 100% glycol methacrylate at − 20 °C for 45 min each.

(7) Infiltrate tissue with 2 changes of embedding monomer at − 20 °C for 45 min each.

(8) Place tissue in capsules in partially polymerized embedding medium at − 20 °C.

(9) Polymerize embedding medium by UV irradiation at − 10 °C to − 12 °C for 48 hr.

The embedding medium consisted of a mixture of glycol methacrylate and *n*-butyl methacrylate, or glycol methacrylate and styrene, with benzoyl peroxide as catalyst.

It was hoped that this low-temperature procedure would reduce extraction during dehydration and embedding, but Cope and Williams (1968) found that the retention of neutral glycerides was no better than with conventional procedures. However, the loss of phosphatidyl choline was reduced (Cope and Williams 1969).

## 6.2   Freeze-drying

In the freeze-drying procedure the ice is removed from the frozen specimen under vacuum. Considerable damage is caused when the freeze-dried tissue is subsequently infiltrated with the liquid embedding medium, and consequently this method of preparation has generally been superceded by freeze-substitution or cryo-ultramicrotomy. Details of the freeze-drying method are given by Sjöstrand (1967) and Rebhun (1972).

## 6.3   Freeze-substitution

The damage caused during the infiltration of dry tissues with the embedding medium can be avoided by using the method of freeze-substitution. Rapid freezing of the tissue is followed by solution ('substitution') of the ice at temperatures below 0 °C (Feder and Sidman 1958). The frozen tissue is kept in the substituting fluid for a long period (Table 6.1) so that the ice dissolves slowly without distorting the structure of the tissue. After removal of the ice the specimen is infiltrated with an embedding medium, often at low temperature. The embedding medium is polymerized in the cold by UV irradiation or by heat in the usual way.

The freeze-substitution methods used in electron microscopy are listed in Table 6.1. Reviews of the development of the technique have been published recently by Bullivant (1970) and Pease (1973), while full practical details are given by Rebhun (1972).

The most important step in the procedure is the initial cryofixation. Ice crystals form during freezing and it is common practice to use a cryoprotective agent, such as glycerol or dimethyl sulphoxide. It must be remembered, however, that these agents may themselves have an effect on the structure of the specimen (Rebhun 1972). For example, dimethyl sulphoxide induces a large increase in volume of the mitotic apparatus in living marine eggs (Rebhun and Sawada 1969). Ice crystal formation can also be reduced by partial dehydration of the tissue before freezing by exposing it to hypertonic conditions (Rebhun and Sander 1971).

The specimen should be cooled as rapidly as possible to keep ice crystal formation to a minimum. Fernández-Morán (1960, 1961) originally proposed using liquid helium II at − 272 °C, but the transfer of heat in liquid helium is poor due to the presence of a film of helium gas which surrounds the specimen (Bullivant 1965). Propane, propylene and Genetron 23 are the best coolants (Rebhun 1972), but Freon 22 is preferred because it is safer to use (Bullivant 1970), and can be easily stored in a deep freeze. It can be re-used after filtration at low temperature through filter paper (Rebhun 1972).

The specimen should be immersed in the coolant as rapidly as possible to prevent dehydration during transfer. Rebhun (1972) describes a simple device for shooting the specimen into the quenching fluid and recommends coating the specimen with a formvar film to reduce evaporation. Monroe et al. (1968) developed an original technique in which small tubular stainless steel projectiles were accelerated to supersonic velocities (400 m/sec) by the explosion of gunpowder. These projectiles passed through the tissue (left ventricle of the heart) cutting out a cylindrical biopsy on the way, and were then collected in a chamber containing liquid propane at − 188 °C (Fig. 6.1).

Rapid freezing can also be obtained by bringing the specimen into contact with the polished surface of a metal block which has previously been cooled to about − 200 °C by immersion in liquid nitrogen (Van Harreveld and

TABLE 6.1

Freeze-substitution methods in electron microscopy

| Pre-treatment of tissue | Coolant | Substitution | | | Embedding | | Reference |
|---|---|---|---|---|---|---|---|
| | | Fluid | Temp. | Time | Medium | Polymerization | |
| Glycerinated | Helium II −272°C | Alcohol-acetone–ethyl chloride | −130°C to −80°C | | Methacrylate −100°C to −75°C | UV −80°C to −20°C | Fernández-Morán (1960, 1961) |
| None | Helium II −272°C | Methanol | −75°C | 2 weeks | Methacrylate | UV −5°C | Bullivant (1960) |
| Glycerinated | Propane* −175°C | Methanol | −75°C | 2 weeks | Methacrylate | UV −5°C | Bullivant (1962) |
| Glycerinated | Propane* −175°C | Ethanol | −75°C | 2 weeks | Methacrylate or Durcupan or Epon | UV −25°C / Heat | Bullivant (1965) |
| None | Freon 12* −150°C | 1% $OsO_4$ in acetone | −80°C | 3 days to 2 weeks | Araldite 502 | Heat | Rebhun (1961) |
| Glycerinated | Propane* −170°C to −160°C | acetone–ethanol (1:1) | −108°C | 2 weeks | Araldite 502 | Heat | Rebhun and Gagné (1962) |
| Hypertonic sea water | Freon 12* or −155°C to −150°C | 1% $OsO_4$ in acetone–ethanol | −100°C | 2 weeks | Epon | Heat | Rebhun and Sander (1971) |

| | | | | | | | | |
|---|---|---|---|---|---|---|---|---|
| None* | Isopentane* | — 160°C | GMA | — 60°C to — 40°C | 2 weeks | GMA | UV — 50° to — 40°C | Bartl (1962) |
| None | Silver* | — 207°C | 2% OsO₄ in acetone | — 85°C | 2 to 3 days | Maraglas | Heat | Van Harreveld and Crowell (1964) |
| None or glycerinated | 70% ethylene glycol or Brass* or Freon 22* | — 77°C — 196°C — 160°C | 70% ethylene glycol in salt solution or Propylene glycol | — 50°C — 40°C | Few hr Few hr | HPMA | | Pease (1967a, b) |
| DMSO | 20% DMSO | — 20°C | Ethanol-25% glutaraldehyde | — 20°C | 2 days | GMA | UV — 12°C to — 10°C | Cope (1968) |
| None | Propane* | — 188°C | Ethanol | — 75°C | 2 weeks | Epon | Heat | Monroe et al. (1968) |

*Key:*

| | |
|---|---|
| * | Cooled with liquid nitrogen |
| UV | Ultra-violet irradiation |
| GMA | Glycol methacrylate |
| HPMA | Hydroxypropyl methacrylate |
| OsO₄ | Osmium tetroxide |
| DMSO | Dimethyl sulphoxide |

Fig. 6.1.   Schematic diagram of a ballistic technique for rapid transfer of a specimen into
a coolant (from Monroe et al. 1968).

Crowell 1964; Pease 1967b). Even with the most rapid freezing methods only a narrow surface layer of a block of tissue is usually free of ice crystals (Van Harreveld and Crowell 1964).

After cryofixation the specimens are passed through the chosen substituting fluids, care being taken to ensure that the fluids are completely dry. Different fluids give very similar results (Rebhun and Sander 1971; Rebhun 1972). One of the main disadvantages of the freeze-substitution method is the long period required in the substituting fluid (Table 6.1). Periods of days or even weeks are required. Pease (1967a) used much shorter times (a few hours) for substitution with ethylene glycol, but it has been suggested by Bullivant (1970) that the tissue was allowed to warm up before substitution was complete. Pease (1973) has recently rejected this criticism.

The tissues may be stained by adding heavy metal salts to the substituting fluid (Fernández-Morán 1960, 1961; Rebhun 1961; Van Harreveld and Crowell 1964; Pease 1967a; Rebhun and Sander 1971), although no staining is observed with osmium tetroxide unless the fluid is allowed to warm up (Bullivant 1970). Alternatively, the sections are stained in the usual way.

The appearance of the tissue is very different from that observed after conventional fixation and embedding. In particular, membranes appear in negative contrast (Fig. 6.2). Even with the 'best' specimens only a small proportion of the cells is free of damage due to ice crystals.

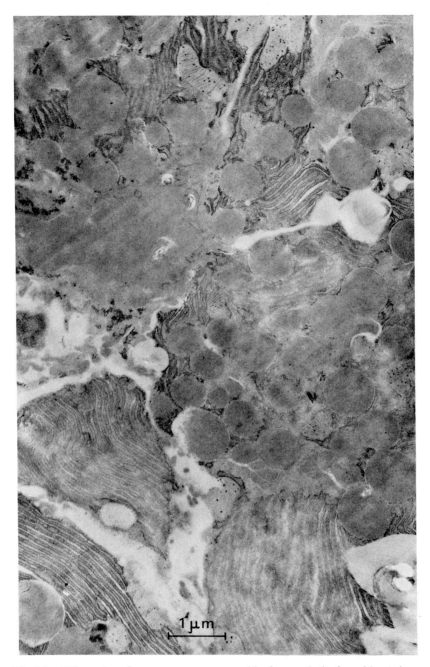

Fig. 6.2. Thin section of mouse pancreas prepared by freeze-substitution without chemical fixation. Embedded in Epon; section stained with lead hydroxide. (Unpublished micrograph from a study by Stanley Bullivant.)

Fig. 6.3.   Ultra-thin dry frozen section of unfixed, unembedded mouse pancreas. The nucleus has light (a) and dense (b) areas. The ribosomes are visible on the rough endoplasmic reticulum (er). The light areas (v) are secretory granules or vesicles.
(From Appleton 1972).

## 6.4 Cryo-ultramicrotomy

In recent years considerable advances have been made in the development of methods for cutting ultrathin sections of frozen specimens (Reid 1974) and special attachments for ultramicrotomes are commercially available. In some methods the tissue is fixed and then embedded in a water-soluble medium, such as gelatin, before freezing and sectioning (e.g. Bernhard and Leduc 1967), while in others the specimens are unfixed and unembedded (e.g. Appleton 1972; Fig. 6.3). This second, technically much more difficult method is necessary in studies on diffusible substances and promises to be of great value in cytochemistry.

The technique is being actively developed in a number of laboratories and it seems likely that before long it will become one of the established techniques in electron microscopy and warrant fuller description in later editions of this book.

REFERENCES

Appleton, T. C. (1972), 'Dry' ultra-thin frozen sections for electron microscopy and X-ray microanalysis: the cryostat approach, Micron *3*, 101.
Bartl, P. (1962), Freeze-substitution method using a water-miscible embedding medium, Proc. 5th Int. Congr. Electron Microscopy, Philadelphia *2*, P-4.
Bernhard, W. and E. H. Leduc (1967), Ultrathin frozen section. I. Methods and ultra-structural preservation, J. Cell Biol. *34*, 757.
Bullivant, S. (1960), The staining of thin sections of mouse pancreas prepared by the Fernández-Morán helium II freeze-substitution method, J. biophys. biochem. Cytol. *8*, 639.
Bullivant, S. (1962), Consideration of membranes and associated structures after cryo-fixation, Proc. 5th Int. Congr. Electron Microscopy, Philadelphia *2*, R-2.
Bullivant, S. (1965), Freeze substitution and supporting techniques, Lab. Invest. *14*, 1178.
Bullivant, S. (1970), Present status of freezing techniques, in: Some biological techniques in electron microscopy, D. F. Parsons, ed. (Academic Press, New York and London), p. 101.
Cope, G. H. (1968), Low-temperature embedding in water-miscible methacrylates after treatment with antifreezes, Jl R. microsc. Soc. *88*, 235.
Cope, G. H. and M. A. Williams (1968), Quantitative studies on neutral lipid preservation in electron microscopy, Jl R. microsc. Soc. *88*, 259.
Cope, G. H. and M. A. Williams (1969), Quantitative studies on the preservation of choline and ethanolamine phosphatides during tissue preparation for electron microscopy. II. Other preparative methods, J. Microscopy *90*, 47.
Feder, N. and R. L. Sidman (1958), Methods and principles of fixation by freeze-substitution, J. biophys. biochem. Cytol. *4*, 593.
Fernández-Morán, H. (1960), Low-temperature preparation techniques for electron microscopy of biological specimens based on rapid freezing with liquid helium II, Ann. N.Y. Acad. Sci. *85*, 689.

Fernández-Morán, H. (1961), The fine structure of vertebrate and invertebrate photo-receptors as revealed by low-temperature electron microscopy, in: The structure of the eye, G. K. Smelser, ed. (Academic Press, New York), p. 521.

Monroe, R. G., W. J. Gamble, C. G. La Farge, R. Gamboa, C. L. Morgan, A. Rosenthal and S. Bullivant (1968), Myocardial ultrastructure in systole and diastole using ballistic cryofixation, J. Ultrastruct. Res. *22*, 22.

Pease, D. C. (1967a), Eutectic ethylene glycol and pure propylene glycol as substituting media for the dehydration of frozen tissue, J. Ultrastruct. Res. *21*, 75.

Pease, D. C. (1967b), The preservation of tissue fine structure during rapid freezing, J. Ultrastruct. Res. *21*, 98.

Pease, D. C. (1973), Substitution techniques, in: Advanced techniques in biological electron microscopy, J. K. Koehler, ed. (Springer-Verlag, Berlin), p. 35.

Rebhun, L. I. (1961), Applications of freeze-substitution to electron microscope studies of invertebrate oocytes, J. biophys. biochem. Cytol. *9*, 785.

Rebhun, L. I. (1972), Freeze-substitution and freeze-drying, in: Principles and techniques of electron microscopy. Biological applications, Vol. 2, M. A. Hayat, ed. (Van Nostrand Reinhold, New York), p. 1.

Rebhun, L. I. and H. T. Gagné (1962), Some aspects of freeze-substitution in electron microscopy, Proc. 5th Int. Congr. Electron Microscopy, Philadelphia *2*, L-2.

Rebhun, L. I. and G. Sander (1971), Electron microscope studies of frozen-substituted marine eggs. 1. Conditions for avoidance of intracellular ice crystallization, Am. J. Anat. *130*, 1.

Rebhun, L. I. and N. Sawada (1969), Augmentation and dispersion of the *in vivo* mitotic apparatus of living marine eggs, Protoplasma *68*, 1.

Reid, N. (1974), Ultramicrotomy, in: Practical methods in electron microscopy, A. M. Glauert, ed. (North-Holland, Amsterdam).

Sjöstrand, F. S. (1967), Electron microscopy of cells and tissues. Vol. 1. Instrumentation and techniques, (Academic Press, New York and London).

Van Harreveld, A. and J. Crowell (1964), Electron microscopy after rapid freezing on a metal surface and substitution fixation, Anat. Rec. *149*, 381.

# Appendix

# List of suppliers

Commercial suppliers of equipment
and materials for fixation, dehydration and embedding.

N.B. The following list includes the names of suppliers known to the author at the present time; she will be interested to hear of names and addresses of other suppliers for inclusion in later editions.

General suppliers are listed first from whom the majority of the smaller items may be obtained. There follows a more comprehensive list with the items arranged in the same order as they are mentioned in the text.

★ Denotes a manufacturer

## 1. General suppliers of materials and equipment for electron microscopy

(a) **Agar Aids for Electron Microscopy**
(Agar Aids)
(U.K. agents for Ladd)
127a Rye Street
Bishop's Stortford
Hertfordshire
England

(b) **Ernest F. Fullam, Inc.** (EFFA)
P.O. Box 444
Schenectady
New York 12301
U.S.A.

**Graticules Ltd.**
(U.K. agents for EFFA)
Sovereign Way
Tonbridge
Kent
England

**Touzart & Matignon**
(French agents for EFFA)
3, Rue Amyot
75 Paris 5
France

(c) **EMscope Laboratories** (EMscope)
99, North Street
London SW4 OHQ
England

(d) **Ladd Research Industries Inc.** (Ladd)
(U.S. agents for Agar Aids)
P.O. Box 901
Burlington
Vermont 05401
U.S.A.

(e) **Polaron Equipment Ltd.** (Polaron)
(U.K. agents for Polysciences)
60/62, Greenhill Crescent
Holywell Estate
Watford
Hertfordshire
England

(g) **Taab Laboratories** (Taab)
52, Kidmore End Road
Emmer Green
Reading
Berkshire
England

**Ted Pella Co.** (Pelco)
(U.S. Agents for Polaron)
P.O. Box 510
Tustin
California 92680
U.S.A.

**Extech International Corp.** (Extech)
(U.S. agents for Taab)
177, State Street
Boston
Massachusetts 02109
U.S.A.

(f) **Polysciences Inc.** (Polysciences)
(U.S. agents for Polaron)
Paul Valley Industrial Park
Warrington
Pennsylvania 18976
U.S.A.

(h) **E. M. Zairyo-Shya**
308, Yotsuya Sango Building
2, Saemoncho Shinijuku-ku,
Tokyo
Japan

## 2. Buffer solutions (§2.2)

(a) *Phospate buffer*

(d) *Collidine buffer*

**Fisons Scientific Apparatus** (Fisons)
Bishops Meadow Road
Loughborough
Leicestershire LE11 ORG
England

| | |
|---|---|
| Ladd | (*see* 1d) |
| Polaron | (*see* 1e) |
| Polysciences | (*see* 1f) |
| Taab | (*see* 1g) |

(e) *Buffer chemicals*

| | |
|---|---|
| Polysciences | (*See* 1f) |

| | |
|---|---|
| General suppliers | (*see* 1) |
| Fisons | (*see* 2a) |

**G. T. Gurr** (Gurr)
Searle Scientific Services
Coronation Road
Cressex Industrial Estate
High Wycombe
Buckinghamshire
England

(b) *Cacodylate buffer*

| | |
|---|---|
| Polaron | (*see* 1e) |
| Taab | (*see* 1g) |

(c) *Veronal-acetate buffer*

(f) *Purified collidine (distilled)*

**Eastman Kodak**
Distillation Products
New York, N.Y.
U.S.A.

| | |
|---|---|
| Ladd | (*see* 1d) |
| Polaron | (*see* 1e) |
| Polysciences | (*see* 1f) |

# 3. Fixatives (Chapter 2)

(a) *Osmium tetroxide* (§ 2.3)

| | |
|---|---|
| General Suppliers | (*see* 1) |
| Fisons | (*see* 2a) |
| Gurr | (*see* 2e) |

**Johnson, Matthey Chemicals Ltd.**
73–83, Hatton Garden
London EC1
England

*Note:* Osmium tetroxide may be returned to Johnson, Matthey Chemicals for regeneration and re-use.

(b) *Glutaraldehyde* (§ 2.4)

| | |
|---|---|
| General suppliers | (*see* 1) |

**Fisher Scientific** (Fisher)
633, Greenwich Street
New York, N.Y. 10014
U.S.A.

| | |
|---|---|
| Gurr | (*see* 2e) |

**Kodak Ltd.**
(U.K. agents for Fisher)
Chemical Division
Kirkby
Liverpool
England

**Union Carbide Corp.**
270, Park Avenue
New York, N.Y. 10017
U.S.A.

(c) *Formaldehyde* (§ 2.5)

| | |
|---|---|
| EFFA | (*see* 1b) |
| EMscope | (*see* 1c) |
| Polaron | (*see* 1e) |
| Polysciences | (*see* 1f) |
| Taab | (*see* 1g) |
| Fisons | (*see* 2a) |
| Gurr | (*see* 2e) |

(d) *Paraformaldehyde* (§ 2.5)

| | |
|---|---|
| EMscope | (*see* 1c) |
| Ladd | (*see* 1d) |
| Polaron | (*see* 1e) |
| Taab | (*see* 1g) |
| Fisons | (*see* 2a) |
| Gurr | (*see* 2e) |

(e) *Acrolein* (acrylaldehyde) (§ 2.6)

| | |
|---|---|
| General suppliers | (*see* 1) |
| Fisons | (*see* 2a) |

**Shell Chemicals Ltd.**
Shell House
Downstream Buildings
London SE1
England

*or:*

**Shell Chemical Corp.**
Industrial Chemical Division
415 Madison Avenue
New York, N.Y.
U.S.A.

(f) *Other aldehydes* (§ 2.6)

| | |
|---|---|
| EMscope | (*see* 1c) |
| Polaron | (*see* 1e) |
| Polysciences | (*see* 1f) |
| Taab | (*see* 1g) |
| Fisons | (*see* 2a) |
| Gurr | (*see* 2e) |

(g) *Potassium permanganate* (§ 2.7)

| | |
|---|---|
| General suppliers | (*see* 1) |
| Fisons | (*see* 2a) |
| Gurr | (*see* 2e) |

(h) *Potassium ferricyanide* (§ 2.8)

| | |
|---|---|
| EMscope | (*see* 1c) |
| Fisons | (*see* 2a) |
| Taab | (*see* 1g) |

(i) *Uranyl acetate* (§ 2.9)

| | |
|---|---|
| General suppliers | (*see* 1) |
| Fisons | (*see* 2a) |
| Gurr | (*see* 2e) |

## 4. Equipment for fixation (Chapter 3)

(a) *Glass vials*, with plastic caps (§ 3.1)

| | |
|---|---|
| Agar Aids | (*see* 1a) |
| EFFA | (*see* 1b) |
| EMscope | (*see* 1c) |
| Polaron | (*see* 1e) |

(b) *Tweezers* (§ 3.1)

| | |
|---|---|
| General suppliers | (*see* 1) |

**Mason & Morton Ltd.**
Fir Tree House
Headstone Drive
Wealdstone, Harrow
Middlesex HA3 5QS
England

(c) *Dental wax* (§ 3.3)

| | |
|---|---|
| EFFA | (*see* 1b) |
| EMscope | (*see* 1c) |
| Ladd | (*see* 1d) |
| Polaron | (*see* 1e) |
| Taab | (*see* 1g) |

(d) *Razor blades*, single-edged (§ 3.3)

| | |
|---|---|
| EFFA | (*see* 1b) |
| EMscope | (*see* 1c) |
| Polaron | (*see* 1e) |
| Taab | (*see* 1g) |

(e) *Millipore membrane filters* (§ 3.5)

**Millipore Filter Corp.**
Millipore House
Abbey Road
London NW10 7SP
England

(f) *Microfuge* (§ 3.8)

**Beckman-RIIC Ltd.**
Eastfield Industrial Estate
Glenrothes
Fife
Scotland

(g) *Centrifuge tubes modified to hold BEEM capsules* (§ 3.8)

| | |
|---|---|
| EFFA | (*see* 1b) |

## 5. Additives for fixation by perfusion (§ 3.3)

(a) *Dextran*

| | |
|---|---|
| Fisons | (*see* 2a) |
| Gurr | (*see* 2e) |
| Polysciences | (*see* 1f) |
| Taab | (*see* 1g) |

(b) *Gum acacia*

| | |
|---|---|
| Fisons | (*see* 2a) |

(c) *Polyvinylpyrrolidone (PVP)*

| | |
|---|---|
| Fisons | (*see* 2a) |
| Gurr | (*see* 2e) |
| Polysciences | (*see* 1f) |
| Taab | (*see* 1g) |

## 6. Encapsulating media (§3.6)

(a) *Agar*

| | |
|---|---|
| Fisons | (*see* 2a) |
| Gurr | (*see* 2e) |
| Polysciences | (*see* 1f) |
| Taab | (*see* 1g) |

(b) *Fibrinogen and thrombin*

**Parke Davis and Co. Ltd.**
Hounslow
Middlesex
England

(c) *Bovine serum albumin* (BSA)

| | |
|---|---|
| EFFA | (*see* 1b) |
| Gurr | (*see* 2e) |
| Polysciences | (*see* 1f) |
| Taab | (*see* 1g) |

(d) *Egg albumen*

| | |
|---|---|
| Fisons | (*see* 2a) |

## 7. Dehydrating agents and intermediate solvents (Chapter 4)

(a) *Ethylene glycol (ethanediol)*

| | |
|---|---|
| Fisons | (*see* 2a) |

(b) *Polyethylene glycol 200 (Carbowax 200)*

| | |
|---|---|
| Fisons | (*see* 2a) |
| Gurr | (*see* 2e) |
| Polysciences | (*see* 1f) |
| Taab | (*see* 1g) |
| Union Carbide | (*see* 3b) |

(c) *Propylene oxide (1,2-epoxy propane)*

| | |
|---|---|
| General suppliers | (*see* 1) |
| Fisons | (*see* 2a) |
| Gurr | (*see* 2e) |

(d) *Styrene*

| | |
|---|---|
| Polaron | (*see* 1e) |
| Polysciences | (*see* 1f) |
| Taab | (*see* 1g) |

## 8. Rotary shakers and mixers (§5.2)

(a) *Epoxy mixer*

| | |
|---|---|
| EFFA | (*see* 1b) |

(b) *Jurand and Ireland Rotary Shaker*

| | |
|---|---|
| EFFA | (*see* 1b) |

(c) *Penetron Rotary Shaker (Kushida design)*

| | |
|---|---|
| Polysciences | (*see* 1f) |

(d) *Rotamixers*

| | |
|---|---|
| EMscope★ | (*see* 1c) |

(e) *Rotary shakers and mixers VEM-16*

**Sakura Finetechnical Co Ltd.★**
9 Honcho 3-chome, Nihombashi
Chuo-ku
Tokyo 103
Japan

(f) *Sunkay 'Swirling' Shaker*

| | |
|---|---|
| Polaron | (*see* 1e) |

(g) *Taab Rotator*

| | |
|---|---|
| Taab | (*see* 1g) |

(h) *Zacks Teflon Stirrer*

Polaron          (*see* 1e)
Polysciences     (*see* 1f)

## 9. Epoxy resin embedding components (§ 5.3)

(a) *Table A.1*

**R. P. Cargille Laboratories Inc.**
Cedar Grove
New Jersey 07009
U.S.A.

**Ciba-Geigy** (Ciba U.K.)
Duxford
Cambridge
England

**Ciba Co., Inc.** (Ciba U.S.)
Plastics Division
Kimberton
Pennsylvania
U.S.A.

*and*

**Ciba Products Corp.**
Fair Lawn
New Jersey
U.S.A.

**Dow Chemical Co.**
Midland
Michigan 48640
U.S.A.

*and*

**Dow Chemical Europe, S.A.**
Alfred Escher Strasse 39
8207 Zurich
Switzerland

**Fluka AG**
Chemische Fabrik
Buchs SG
Switzerland

(b) *Epok 533*

**Oken Shoji Co.**★
311, Kobikikan 7
6-chome
Ginza-higashi
Chuo-ku
Tokyo
Japan

(c) *Maraglas 655*

**Marblette Corp**★
Long Island City
New York
U.S.A.

(d) *DMP-30*

**Rohm & Haas Co.**
Washington Square
Philadelphia
Pennsylvania
U.S.A.

(e) *BDMA (Benzyl dimethyl amine)*

**Maumee Chemical Co.**
2, Oak Street
Toledo
Ohio
U.S.A.

(f) *DMAE (Dimethylaminoethanol) (S-1)*

**Pennsalt Chemical Corp.**
Three Penn. Center
Philadelphia
Pennsylvania 19102
U.S.A.

# TABLE A.1

Suppliers of epoxy resin embedding components

Key: ○ Manufacturer  w-s water soluble

| | Cargille | Ciba (U.K.) | Ciba (U.S.) | Dow | EFFA (see 1b) | EMscope (see 1c) | Fluka | Gurr (see 2e) | Ladd (see 1d) | Polaron (see 1e) | Polysciences (see 1f) | Shell (see 2a) | Taab (see 1g) | Union Carbide (see 3b) | Zairyo-Shya (see 1h) |
|---|---|---|---|---|---|---|---|---|---|---|---|---|---|---|---|
| **Embedding kits** | | | | | | | | | | | | | | | |
| Araldite CY 212 **or** | × | — | — | — | — | × | — | — | — | × | × | — | × | — | — |
| Durcupan ACM | — | — | — | — | — | × | × | × | — | × | — | — | × | — | — |
| Araldite 502 | × | — | — | — | — | — | — | — | — | × | — | — | — | — | — |
| Araldite 506 | — | — | — | — | × | — | — | — | — | — | — | — | — | — | — |
| Araldite 6005 | — | — | — | — | × | — | — | — | — | × | — | — | — | — | — |
| DER 332 | — | — | — | — | — | — | — | — | — | — | × | — | — | — | — |
| Epon 812 | × | — | — | — | × | × | — | — | × | × | × | × | × | — | — |
| Maraglas 655 | × | — | — | — | × | — | — | — | × | × | × | — | — | — | — |
| ERL 4206 | — | — | — | — | × | × | — | — | × | × | × | × | — | × | — × |
| Durcupan (w–s) | — | — | — | — | — | × | × | × | — | × | × | — | × | — | — |
| **Epoxy resins** | | | | | | | | | | | | | | | |
| Araldite CY 212 **or** | — | ○ | — | — | — | × | — | — | — | — | — | — | × | — | — |
| Durcupan ACM | — | — | — | — | — | × | × | × | — | × | — | — | × | — | — |
| Araldite 502 | × | — | ○ | — | × | — | — | — | × | × | × | — | — | — | — |
| Araldite 506 | × | — | ○ | — | × | — | — | — | × | × | — | — | — | — | — |
| Araldite 6005 | × | — | ○ | — | × | — | — | — | × | × | — | — | — | — | — |
| DER 332 | — | — | — | ○ | — | — | — | — | × | × | — | — | — | — | × |
| DER 334 | — | — | — | ○ | — | — | — | — | — | — | — | — | — | — | — |
| ERL 4206 | — | — | — | — | × | — | — | — | × | × | × | × | — | × | ○ × |
| Epon 812 | — | — | — | — | × | × | — | × | × | × | × | × | ○ | × | — × |
| Epon 815 | — | — | — | — | — | — | — | — | — | × | — | ○ | — | — | × |
| Maraglas 655 | — | — | — | — | × | — | — | × | × | × | × | — | — | — | — |
| Aquon (w-s) | — | — | — | — | — | — | — | — | — | — | — | × | — | — | — |
| Durcupan (w-s) | — | — | ○ | — | — | × | × | × | — | × | — | — | × | — | × |
| **Hardeners** | | | | | | | | | | | | | | | |
| DDSA | — | × | — | — | × | × | × | × | × | × | × | × | — | × | — × |
| HHPA | — | — | — | — | — | — | — | × | — | × | × | — | × | — | — |
| MNA **or** NMA | — | — | — | — | × | × | — | × | × | × | × | — | × | — | × |
| NSA | — | — | — | — | — | × | — | — | — | × | × | — | × | — | × |
| **Accelerators** | | | | | | | | | | | | | | | |
| BDMA | — | — | — | — | × | × | — | × | — | × | × | — | × | — | — |
| DMAE (S–1) | — | — | — | — | — | × | — | — | — | × | × | — | × | — | — |
| DMP-30 | — | × | × | — | × | × | × | × | × | × | × | × | — | × | — |
| **Plasticizers and flexibilizers** | | | | | | | | | | | | | | | |
| Cardolite NC-513 | — | — | — | — | × | — | — | × | × | × | × | — | — | — | — |
| DER 732 | — | — | — | ○ | — | — | — | — | — | × | × | — | — | — | × |
| DER 736 | — | — | — | ○ | — | × | — | — | — | × | × | — | × | × | × |
| Dibutyl phthalate | — | × | × | — | × | × | × | × | — | × | × | — | × | — | — |
| Triallyl cyanurate | — | — | — | — | — | — | — | — | — | — | — | × | — | × | — |

**Fabriek van Chemische Producten**
(European agents for Pennsalt)
Vondelingenplaat N.V.
P.O. Box 7120
Rotterdam
The Netherlands

(g) *Thiokol LP-8*

**Thiokol Chemical Corp.★**
Trenton
New Jersey
U.S.A.

## 10. Polyester resin embedding kits (§ 5.4)

(a) *Selectron*

**Pittsburgh Plate Glass Co.★**
Plastic Sales
Paint and Brush Division
Pittsburgh
Pennsylvania
U.S.A.

Polysciences                    (*see* 1f)

(b) *Vestopal W*

Polaron                         (*see* 1e)
Polysciences                    (*see* 1f)

## 11. Polyester resins (§ 5.4)

(a) *Beetle 4116 and 4134*

**B.I.P. Chemicals Ltd.★**
Oldbury
Birmingham
England

(b) *Rhodester 1108*

**Société des Usines Chimiques★**
Rhône-Poulenc
France

(c) *Rigolac 2004 and 70F*

**Riken Goseijushi Co.★**
3, 6-chome
Ginza 6-3
Chuo-ku
Tokyo
Japan

(d) *Selectron*

Pittsburgh Plate Glass Co.    (*see* 10a)
Polysciences                  (*see* 1f)

(e) *Vestopal W*

**Martin Jaeger★**
Vésenaz
Geneva
Switzerland

Gurr                          (*see* 2e)
Polaron                       (*see* 1e)
Polysciences                  (*see* 1f)
Taab                          (*see* 1g)

## 12. Initiators and accelerators for polyester resins and methacrylates (§ 5.4 and § 5.5)

(a) *Azo-bis-iso-butyronitrile*

**Eastman Organic Chemicals**
Rochester 3
New York 14650
U.S.A.

Polysciences                    (*see* 1f)
Taab                            (*see* 1g)

(b) *Benzoin*

| | |
|---|---|
| Fisons | (*see* 2a) |
| Polysciences | (*see* 1f) |
| Taab | (*see* 1g) |

(c) *Benzoyl peroxide*

| | |
|---|---|
| Fisons | (*see* 2a) |
| Gurr | (*see* 2e) |
| Ladd | (*see* 1d) |
| Polysciences | (*see* 1f) |
| Taab | (*see* 1g) |
| Polaron | (*see* 1e) |

(d) *Benzoyl peroxide paste (Luperco)*

| | |
|---|---|
| Gurr | (*see* 2e) |
| Polysciences | (*see* 1f) |
| Riken Goseijushi★ | (*see* 11c) |
| Taab | (*see* 1g) |

**Wallace and Tiernan Inc.**
Lucido Division
Buffalo
New York 14240
U.S.A.

(e) *Cobalt naphthenate*

| | |
|---|---|
| Gurr | (*see* 2e) |
| Jaeger | (*see* 11e) |
| Polysciences | (*see* 1f) |
| Taab | (*see* 1g) |

(f) *Divinyl benzene*

| | |
|---|---|
| Dow | (*see* 9a) |
| Polysciences | (*see* 1f) |
| Taab | (*see* 1g) |

(g) *t-butyl perbenzoate*

| | |
|---|---|
| Gurr | (*see* 2e) |
| Jaeger | (*see* 11e) |
| Taab | (*see* 1g) |
| Wallace and Tiernan | (*see* 12d) |

(h) *Uranyl nitrate*

| | |
|---|---|
| Fisons | (*see* 2a) |
| Gurr | (*see* 2e) |
| EMscope | (*see* 1c) |
| Ladd | (see 1d) |
| Taab | (*see* 1g) |
| Polaron | (*see* 1e) |

## 13. Methacrylate embedding kits (§ 5.5)

(a) *n-butyl/methyl methacrylate*

| | |
|---|---|
| Ladd | (*see* 1d) |
| Polaron | (*see* 1e) |

(b) *Glycol methacrylate (GMA) (Water-soluble)*

| | |
|---|---|
| Polaron | (*see* 1e) |
| Polysciences | (*see* 1f) |
| Zairyo-Shya | (*see* 1h) |

(c) *HPMA (hydroxypropyl methacrylate)*

| | |
|---|---|
| Polaron | (*see* 1e) |
| Taab | (*see* 1g) |
| Zairyo-Shya | (*see* 1h) |

## 14. Methacrylates (§ 5.5)

(a) *n-butyl and methyl methacrylates*

| | |
|---|---|
| Gurr | (*see* 2e) |
| Ladd | (*see* 1d) |
| Polaron | (*see* 1e) |
| Polysciences | (*see* 1f) |
| Taab | (*see* 1g) |

(b) *Glycol methacrylate (GMA) (2-hydroxy ethyl methacrylate)*

**Eastman Kodak Co.★**
343, State Street
Rochester
New York 14650
U.S.A.

(b) *Glycol methacrylate (continued)*

| Kodak Ltd. | *(see* 3b) |
| Polaron | *(see* 1e) |
| Polysciences | *(see* 1f) |
| Rohm & Haas | *(see* 9d) |
| Taab | *(see* 1g) |

(c) *HPMA (2-hydroxy propyl methacrylate)*

| Polaron | *(see* 1e) |
| Polysciences | *(see* 1f) |
| Rohm & Haas | *(see* 9d) |
| Taab | *(see* 1g) |

(d) *Other methacrylates*

| Polysciences | *(see* 1f) |
| Taab | *(see* 1g) |

## 15. Capsules (Chapter 5)

(a) *Gelatin capsules*

| General Suppliers | *(see* 1) |
| Parke Davis | *(see* 6b) |

(c) *Taab polyethylene and polypropylene capsules*

| Taab | *(see* 1g) |

(d) *Polaron 'peelcap' capsules*

| Polaron ★ | *(see* 1e) |

(b) *BEEM polyethylene capsules*

**Better Equipment for Electron Microscopy Inc. (BEEM)** ★
P.O. Box 132
Jerome Avenue Station
Bronx
New York 10468
U.S.A.

| General suppliers | *(see* 1) |

(e) *BEEM Bojax embedding blocks (sets of plastic capsules)*

| BEEM ★ | *(see* 15b) |
| Ladd | *(see* 1d) |
| Pelco | *(see* 1e) |

(f) *Micromoulds (sets of plastic capsules)*

| EFFA | *(see* 1b) |
| Pelco | *(see* 1e) |

## 16. BEEM capsule holders (§ 5.2)

| BEEM ★ | *(see* 15b) |
| EFFA | *(see* 1b) |
| EMscope | *(see* 1c) |
| Ladd | *(see* 1d) |
| Polaron | *(see* 1e) |

## 17. Embedding moulds for flat embedding (§ 5.8)

(a) *EFFA mini-vials*

| EFFA ★ | *(see* 1b) |

(b) *Polyethylene moulds*

**Nalge Co. Inc.** ★
Rochester
New York
U.S.A.

(c) *Silicone rubber moulds*

| | |
|---|---|
| EFFA | (*see* 1b) |
| Ladd | (*see* 1d) |
| Polaron | (*see* 1e) |

**C. W. French Inc.**
58, Bittersweet Lane
Weston
Massachusetts 02193
U.S.A.

(d) *Vinyl cups*

**Fabri-Kal Co.★**
Kalamazoo
Michigan
U.S.A.

## 18. Ultra-violet lamps (§ 5.2)

(a) *UV curing chamber*

Ladd                    (*see* 1d)

(b) *UV lamps*

| | |
|---|---|
| EFFA | (*see* 1b) |
| Polysciences | (*see* 1f) |
| Taab | (*see* 1g) |
| Polaron | (*see* 1e) |

**P. W. Allen and Co.**
253, Liverpool Road
London W1
England

## 19. Substrates for cell monolayers (§ 3.5 and § 5.8)

(a) *Falcon Petri dishes (polystyrene)*

**Falcon Plastics Co.★**
Culver City
California
U.S.A.

**Bio-Cult Laboratories Ltd.**
3, Washington Road
Abbotsinch Industrial Estate
Paisley
Scotland

(b) *Nunclon Petri dishes*

**Nunc A/S ★**
Roskilde
Denmark

**James A. Jobling and Co. Ltd.**
(U.K. agents for Nunc A/S)
Stone
Staffordshire ST15 OBG
England

(c) *Melinex film (polyester)*

**Boyden Data Papers, Ltd.**
Trade Service Division
Parkhouse Street
Camberwell
London SE5
England

(d) *Mica*

| | |
|---|---|
| Agar Aids | (*see* 1a) |
| EFFA | (*see* 1b) |
| EMscope | (*see* 1c) |
| Polaron | (*see* 1e) |

(e) *Teflon-coated coverslips*

Polysciences            (*see* 1f)

(f) *Nucleopore filters*

**General Electric Company**
U.S.A.

## 20. Dimethyl sulphoxide *(cryo-protective agent)* (§6.1)

| | |
|---|---|
| EMscope | *(see* 1c) |
| Fisons | *(see* 2a) |
| Polysciences | *(see* 1f) |
| Taab | *(see* 1g) |

## 21. Coolants (§6.3)

(a) *Freon 12 (dichlorodifluoromethane) and Freon 22 (monochlorofluoromethane)*

Under trade name '*Can-o-gas*'

**Balzers AG**
FL 9496 Blazers
Liechtenstein

*and*

**Balzers High Vacuum Ltd**
Northbridge Road
Berkhamsted
Hertfordshire
England

*and*

**Balzers High Vacuum Corporation**
PO Box 10816
Santa Ana
California 92711
U.S.A.

**Dean and Wood (London) Ltd**
83/85 Mansell Street,
London, E1
England

Under trade name '*Halocarbon 12 and 22*'

**Air Products**
Speciality Gases Department
Stenor Street
Cobridge, Burslem
Stoke-on-Trent
England

Under trade name '*Ucon*' *(Freon 22)*

**J. T. Baker Chemical Company**
Phillipsburg
New Jersey
U.S.A

Under trade name '*Arcton*'

**I.C.I. Ltd**
Mond Division
Rocksavage Works
Runcorn
Cheshire
England

(b) *Isopentane*

**BDH Chemicals Ltd**
Poole,
Dorset, BH12 4NN
England

(c) *Propane*

BDH (see 21b)

# Index for list of suppliers

Numbers referred to are the sections in the preceding list of addresses

# Subject index

# ULTRAMICROTOMY

NORMA REID

*LKB-Produkter AB*
*Far East Operations*
*Hong Kong*

# Contents of part II

# Acknowledgements

I am deeply grateful to Dr. A. M. Glauert and Mr. A. Persson for their valuable help and constructive criticism in preparing this manuscript and also to Dr. Glauert for her contribution to Chapter 10.

I am indebted to LKB for all the facilities they have put at my disposal and to all those members who have helped with the preparation of this book. In particular Mr. J. Bohman for making all the drawings, Mr. L. Kindle for the micrographs in Figs. 1.3 and 1.4 and Dr. L. Seveus for supplying much of the information on cryo-ultramicrotomy.

My gratitude goes to the following Companies, Du Pont Company (Sorvall Operations), EXAL, Kent Cambridge Medical, Polaron Equipment Ltd and C. Reichert AG for their co-operation and permission to reproduce the photographs of their equipment. Also to Dr. Pedler and the Journal of Microscopy for permission to reproduce Fig. 11.1.

Finally I would like to thank my husband for his understanding and tolerance and without whose encouragement this book would not have been written.

*February 1974* NORMA REID

# Chapter 1

# Introduction

Ultramicrotomy is the term used today to describe the technique of producing sections of material thin enough for examination in the electron microscope.

The majority of materials are too thick in their natural state to be examined directly in the electron microscope due to the limited penetration of the electron beam. Ultramicrotomy provides a solution to this problem and has been used successfully to study all types of plant and animal tissue and other such varied materials as metals, plastics, synthetic fibres and even samples from the moon. Using conventional electron microscopes operating at accelerating voltages up to 100 kV, the sections should be in the range of 10 to 100 nm thick and are termed ultrathin sections.

To be able to cut ultrathin sections it is necessary for the material to be fairly hard. Most animal and plant tissues are relatively soft and so must first be embedded in a hard resin. The techniques for this are described in a companion book (Glauert 1974). Naturally occurring hard materials such as metals can often be cut without embedding.

With modern ultramicrotomes it is relatively easy to cut sections 50 to 100 nm thick from the majority of biological specimens. These sections are perfectly adequate for most studies of fine structure and have the advantage over thinner sections that sufficient of the specimen is included in the section to give some idea of the interrelationships of different structures. Sections less than 50 nm thick may be required for high resolution work and this will call for a little more care and patience. Before examination in the electron microscope the sections are normally stained to enhance their contrast; these techniques are described in another volume in this series (Lewis et al. 1974). Before proceeding to electron microscopy it is advisable to have a clear understanding of the structure of the material under study, at the light microscope level. To facilitate the correlation between light and electron microscopy, it is becoming increasingly common to cut sections 0.5 to 2 µm in thickness from the same specimen for examination under the light microscope. These sections are termed thick sections. Ultramicrotomy of non-biological materials is generally a little more difficult and can cause some deformation of the specimen. By comparing the results with those obtained by other thinning methods (Goodhew 1972), it will be seen whether ultramicrotomy is worthwhile in these cases.

The techniques for obtaining ultrathin sections from all types of material are basically the same. It is the object of this book to describe these in sufficient detail, together with the problems that maybe encountered, to enable anyone to perform them successfully. These techniques are now routine in many laboratories and none of them are particularly difficult; all that is required is patience and attention to detail.

An alternative method to embedding, for hardening the material, is freezing and sectioning at low temperatures. This relatively new technique of cryo-ultramicrotomy, which can be applied to biological and non-biological specimens, is still in the development stage and is therefore only briefly described here. Reference to original works are included for those who wish to study the technique further. The literature survey for this book ended in January 1974.

### 1.1   The formation of a section

The production of sections by ultramicrotomy involves a type of cutting action similar to that used in the machining of metals. A prepared specimen is moved past a cutting edge and forces are transmitted from that cutting edge to the specimen causing a thin layer of the material to be detached. Unlike machining the primary concern in ultramicrotomy is with the section and not with the surface finish.

As a piece of material is severed from the face of the specimen block it is strained into a new position from the plane of cutting to the sectioning facet of the knife (Fig. 1.1). In the cutting of metals this stress is relieved by the formation of slip steps (Fig. 1.2). This results in small furrows on both surfaces of the section which can be seen as bands of varying contrast in the electron microscope (Phillips 1962). Considerable friction is encountered as the section moves down the sectioning facet of the knife but this becomes almost negligible if a suitable liquid such as water is present at the

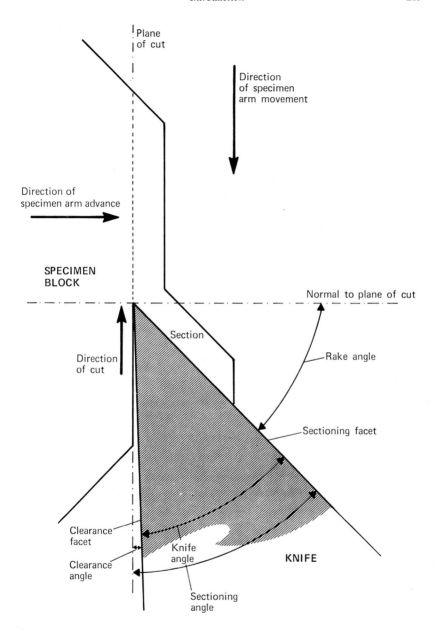

Fig. 1.1. Diagram illustrating the directions and angles involved in sectioning. The sectioning angle is made up of the clearance angle and the knife angle. The sectioning angle, together with the rake angle, form an angle of 90°.

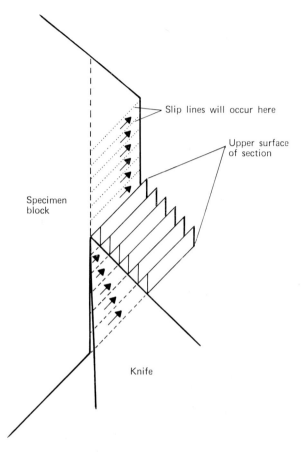

Fig. 1.2.    When cutting metals the section may consist of a series of chips, each of which has slipped a little in relation to its neighbour. This results in a corrugated surface on both sides of the section, and is seen as bands, varying in contrast, in the electron microscope.

knife edge; the section then floats easily onto the liquid surface. When cutting a more elastic material the strain can be absorbed without permanent deformation. The section is bent as it hits the knife edge but it will straighten out if it is allowed to float freely on a liquid surface (Figs. 1.3 and 1.4). In the ideal situation the section would be exactly the same size and shape as the face of the specimen block from which it was detached, with a thickness equal to the advance of the specimen towards the knife. However, in practice some deformation remains resulting in a shortening of the section in the direction of cutting together with an increase in the thickness. This is com-

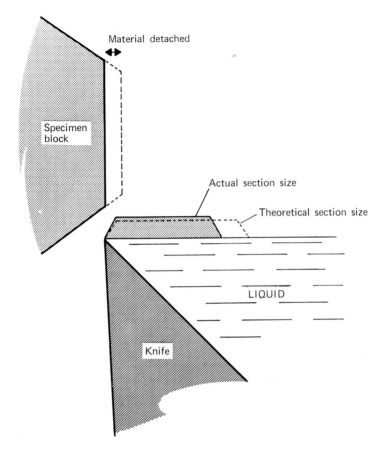

Fig. 1.3. The actual section is always shorter than the block face, although the width remains the same. This results in the section being thicker than the piece of material detached from the specimen block.

monly referred to as compression although it is really a redistribution of the material (Fig. 1.3).

In ultramicrotomy, as in both machining and normal microtomy, variations of the knife angle and clearance angle will affect the amount of stress introduced into the material and thus vary the quality of the results. Other factors influencing the quality of the sections are the speed of the cut, the temperature, and the sharpness of the cutting edge. The production of good sections is dependent on finding the optimum conditions for the material under observation.

Fig. 1.4.    Sections adhering together in ribbon formation while floating
on a liquid surface.

## 1.2    The need for serial sections

If only single sections are examined in the electron microscope interpretation
of the structure may be inaccurate. Artefacts can be introduced during
the cutting process or in the subsequent handling of the section and could
provide misleading information if only one section is examined. Furthermore
the relationship between various structures cannot be accurately judged

from a single section because it only provides information in two dimensions. Structures that appear in one section as two separate profiles may in reality result from the branching at another level of a single structure (e.g. Pedler and Tilly 1968). By studying several sections cut in serial order it is possible to interpret the third dimension. As the sections are produced they will adhere to one another along one edge forming a 'ribbon' (Fig. 1.4). If the ribbon is transferred intact to the electron microscope the serial sections can be studied in sequence and will provide more useful and reliable information than sections studied singly.

REFERENCES

Glauert, A. M. (1974), Fixation, dehydration and embedding of biological specimens, in: Practical methods in electron microscopy, A. M. Glauert, ed. (North-Holland, Amsterdam).

Goodhew, P. F. (1972), Specimen preparation in materials science, in: Practical methods in electron microscopy, A. M. Glauert, ed. (North-Holland, Amsterdam).

Lewis, P. R., D. P. Knight and M. A. Williams (1974), Staining methods for thin sections, in: Practical methods in electron microscopy, A. M. Glauert, ed. (North-Holland, Amsterdam).

Pedler, C. and R. Tilly (1968), A source of error in the interpretation of electron micrograph, Jl R. microsc. Soc. *88*, 177.

Phillips, V. A. (1962), The deformation of metals and alloys by ultramicrotomy, in: Direct observation of imperfections in crystals, J. B. Newkirk and J. H. Wernick, eds. (J. Wiley and Sons, New York and London), p. 179.

# Chapter 2

# Ultramicrotomes

## 2.1 Historical survey

Attempts to obtain sections thin enough for the electron microscope were first made by Von Aldenne (1939). A Minot rotary microtome was modified to produce wedge-shaped sections and the electron beam was able to penetrate the thinner edges of the sections. This was followed by several ultramicrotomes of various designs made in different laboratories. One theory put forward the idea that the shorter the time the knife is in contact with the specimen block the lesser would be the strain put on the section. This led to the design of two high speed microtomes using a razor blade as the cutting edge rotating at very high speeds past the specimen block (O'Brien and McKinley 1943; Fullam and Gessler 1946). Sections were thrown off in all directions making it very difficult to collect them. This idea was abandoned when improvements in embedding techniques enabled slower cutting speeds to be used and led to new designs of instruments.

When the advantages of floating the sections on a liquid surface were realized it became necessary to keep the knife stationary, to avoid movement of the liquid, whilst the specimen block was moved past it in the vertical plane. A section is detached during the downward stroke but if the specimen block is returned upward along the same path the section is immediately picked up by the specimen block as it passes the knife edge. The reason for this is that there is no clearance between the knife edge and the specimen block so that the liquid from the knife flows onto the face of the specimen block carrying the section with it. To avoid this problem 'single pass' ultramicrotomes were developed in which the specimen block passes the knife edge only once, on the downward cutting stroke, returning via another path.

Further developments in embedding and the introduction of glass knives by Latta and Hartmann (1950) led to improvements in design and the production of commercial instruments. A comprehensive review of early ultramicrotomes is given by Gettner and Ornstein (1956); only those in common usage today will be described here.

## 2.2 Ultramicrotome design

Certain functions are common to all modern ultramicrotomes although they may be performed in different ways. An arm carrying the specimen block is moved downwards past a sharp cutting edge, the knife, and a very small movement of the specimen arm towards the knife between each cut results in a thin slice or section being detached from the face of the specimen block. A small trough is attached to the knife to hold the liquid on which the sections will float. All ultramicrotomes have three ranges of advance, a coarse control for initial alignment, a fine advance for cutting sections up to several micrometres thick for light microscopy, and an ultra-fine advance for ultrathin sections. The coarse control is made by moving the knife towards the specimen block but for the ultra-fine advance it is usually the specimen arm that is moved, towards the knife. Either system can be used to produce the intermediate or fine advance. The ultra-fine advance is achieved either by mechanical means or by thermal expansion, but the common objective is an accurate linear movement which will remain reproducible during the life of the instrument. Once a section has been cut it normally remains attached to the knife edge and, as discussed earlier, the specimen block should not be allowed to pass close to the knife on its upward return stroke or the section will be removed. Avoiding action is taken by deflecting the specimen arm sideways (Fig. 2.1), or backwards (Fig. 2.2), or by retracting the knife (Fig. 2.3), during the return stroke.

The cutting speed is the rate at which the specimen passes the knife during the cutting stroke and is generally expressed in millimetres per second. This speed should remain constant over a specified distance. A selection of cutting speeds is required to cope with the various different consistencies and sectioning properties of different specimens. The cutting force is provided by gravity or by an applied force such as a motor. Different cutting speeds are obtained when using the effect of gravity by applying a variable braking force, and with a motor by varying its speed. The remainder of the cycle, the return stroke, during which the arm carrying the specimen is returned to its starting position, is performed by employing a motor or the arm is lifted

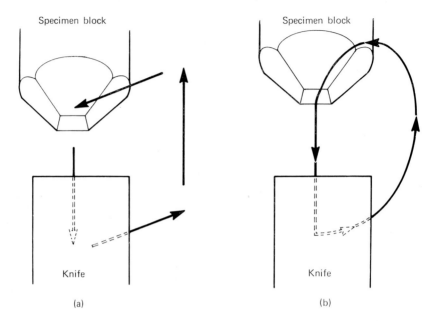

Fig. 2.1.   Movement of the specimen block in relation to the knife. The broken lines indicate that part of the stroke which is behind the knife when viewed from the front. (a) The specimen arm is deflected sideways so that the specimen block moves through a parallelogram-shaped path. (b) The specimen block moves through a D-shaped path, the downward cutting stroke forming the straight leg.

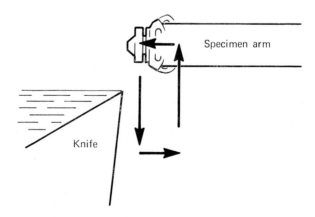

Fig. 2.2.   The specimen arm of the Sorvall Porter-Blum MT2 ultramicrotome is moved backwards from the knife edge during the return stroke (side view).

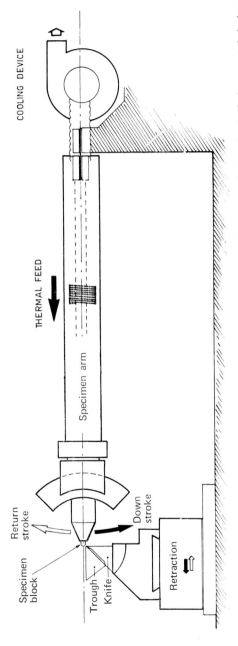

Fig. 2.3. The specimen arm in the **LKB** Ultrotomes is moved up and down along the same path, but the knife is retracted during the upward stroke until the specimen block has cleared the knife edge (side view). (By courtesy of **LKB Produkter AB.**)

manually. This part of the cycle is usually accomplished as quickly as possible to have the minimum delay between cutting strokes.

During the cutting process the specimen, the knife and the ultramicrotome are all placed under stress. This causes distortions within them which will relax in various periods of time. The front of the specimen block is compressed by the cutting action and, when released from contact with the knife, will expand, the amount of expansion being dependent on the interval of time that elapses before its next contact with the knife. Contact with the specimen block causes the tip of the knife to vibrate slightly and if the subsequent cut is made too quickly these vibrations may not have died out. Variations in temperature in the region of the knife and specimen arm are very critical as they will cause expansion or retraction of various parts of the ultramicrotome. Temperature disturbances can be caused by the operator holding parts of the instrument or breathing heavily over the cutting area. All these influences will affect the reproducibility of the thickness of the sections and are therefore undesirable. A regulated stroke movement will help to keep these effects constant and therefore produce more uniform serial sections (Porter 1964). It is for this reason that today most ultramicrotomes have an automated specimen arm movement, although they can be operated manually as well. An additional advantage of automatic operation is that the operator does not have to touch the instrument during sectioning which avoids the risk of introducing vibrations by such contact.

Random movements within the ultramicrotome caused by friction, distortion or thermal changes will have an adverse effect on the reproducible performance of the instrument (Gettner and Ornstein 1956). Care should be taken to follow the manufacturer's instructions in the use and maintenance of the instrument in order to keep these random movements to a minimum.

Both glass and diamond knives can be used with all the ultramicrotomes described below, although certain instruments may require special holders for diamond knives. Specimen holders are available to take various sizes of specimen block up to 8 mm in diameter and special vice-type holders are used to accommodate flat-embedded specimens. It is important to select the correct sized holder for each specimen block or it may not be securely clamped, with the result that the block will move out of position when it hits the knife edge and the correct thickness of section will not be obtained.

A binocular microscope together with suitable lighting is required to observe the sections during the cutting process and afterwards when collecting them. The magnification of the microscope should cover a range of at least $20 \times$ to $40 \times$ as collection of the sections can be made more easily using lower magnifications but higher magnifications will be needed to see if the sections are free from faults. The lighting is normally provided by a fluorescent lamp, as this does not produce too much heat and, when reflected off the surface of water, provides a white background against which variations in thickness of the sections can be seen as different colours. Critical examination of the knife edge cannot be made with this diffuse lighting and an additional focused light source is needed to see the imperfections in the knife edge.

### 2.3 Cambridge Huxley Mark 1 and Cambridge Huxley Mark 2 (or LKB Huxley) ultramicrotomes (Figs. 2.4-2.6)

The first instrument, Mark 1, (Fig. 2.4) was designed by Professor A. F. Huxley over fourteen years ago (Huxley 1957) but it is still in use in many laboratories today. It is manually operated and employs a mechanical advance. The specimen arm is connected by an adjustable wire suspension to the operating lever and is hinged by flexible steel strips to the advance mechanism. A micrometer spindle advances the specimen arm via a 250:1 reduction lever in steps of 5 nm, from 0 to 150 nm, automatically between each stroke. The arm is raised to the top of the cutting stroke by lifting the operating lever and is then allowed to drop under the action of gravity. An oil-filled dash-pot controls the rate of fall past the knife edge. The cutting speed can be varied by adjusting a needle valve on the piston of the dash-pot to give a small range of speeds although these are not calibrated. The specimen arm is displaced sideways during its return stroke as in Fig. 2.1b.

The base of the knife rests on an adjustable screw and this determines the height of the knife in the knife holder. The specimen arm can be arrested in the middle of its stroke by the introduction of a supporting bar and then the specimen arm, and consequently the cutting stroke, can be raised or lowered by adjusting the wire suspension until the specimen block rests at the same height as the knife edge. The knife can be tilted, rotated and moved laterally by movements of the knife stage, and a micrometer adjustment provides a fine advance movement of the knife stage towards the specimen arm. The standard specimen block holder does not allow for any orientation but a special orientating head is available. The binocular microscope and lamp are fitted on a separate stand placed by the side of the instrument.

This model is no longer available since a new version, the Mark 2, was introduced in 1970 (Figs. 2.5 and 2.6). Although manufactured by Kent Cambridge Medical it is only marketed by them in the U.K.; throughout

Fig. 2.4. The Cambridge Huxley Mark 1 ultramicrotome. The fibre glass protective cover is not shown. (a) 250:1 reduction lever; (b) flexible steel strips; (c) wire suspension; (d) oil dashpot; (e) specimen arm; (f) specimen block holder; (g) knife holder; (h) fine advance; (i) micrometer spindle; (j) operating lever. (By courtesy of Kent Cambridge Medical.)

Fig. 2.5. The Cambridge Huxley Mark 2 (or LKB Huxley) ultramicrotome. The small, hand-size motor control box is not shown. The microscope stand rests on the table and can be moved backwards and forwards. Either an Olympus (as shown) or a Bausch and Lomb stereomicroscope can be fitted. (By courtesy of Kent Cambridge Medical.)

the rest of the world it is marketed through the LKB organisation and is also known as the LKB Huxley ultramicrotome. The working principles are the same as in the Mark 1 but a motor has been fitted in the base of the instrument so that the specimen arm can be operated automatically or in single strokes. The motor lifts the specimen arm to the top of its stroke and is then disconnected during the downward cutting stroke to avoid vibrations. Adjustments to the cutting speed can be made much more easily than in the Mark 1, by means of an external control, although the speeds are

Fig. 2.6.    The interior of the Cambridge Huxley Mark 2 (or LKB Huxley) ultramicrotome.
(a) 250:1 reduction lever; (b) flexible steel strips; (c) specimen arm; (d) oil dash-pot; (e)
speed control; (f) polypropylene suspension; (g) knife holder; (h) macro feed; (i) micro
feed; (j) manual operation lever; (k) selector knob. (By courtesy of
Kent Cambridge Medical).

still not calibrated. The movements of the knife stage have been improved
and a Macro feed control has been added to the specimen arm advance.
When the selector knob is turned to 'Macro' the normal ultra-fine advance
mechanism is disconnected. A manually operated advance of 0.5 μm, 1 μm
or 2 μm can then be made between each stroke. With the selector knob in the
'Micro' position only the ultra-fine advance is operational. A much improved
microscope stand fitting around the front of the instrument carries the binocular microscope and fluorescent lamp on two arms, both of which are
moveable through an arc around the cutting point.

## 2.4   Sorvall Porter–Blum MT1 and MT2 ultramicrotomes (Figs. 2.7–2.9)

The first ultramicrotome developed by Porter and Blum (1953) was a hand-

Fig. 2.7. The Sorvall 'Porter-Blum' MT1 ultramicrotome. (By courtesy of
Du Pont Company, Sorvall Operations.)

operated single-pass instrument using a 60 W tungsten lamp placed above
the specimen arm to provide a thermal advance. It depended on the skill of
the operator to judge when the arm had advanced sufficiently to make
another stroke. A mechanical advance system was introduced in a later model
and this then became the basis of the first commercial MT1 ultramicrotome
(Fig. 2.7).

A specimen block is fixed to one end of a cantilevered arm, the other end of
which is mounted on off-set pivots so that the arm is free to move both
horizontally and vertically. The end of the arm carrying the specimen block
follows a guided path and is displaced sideways during the return stroke as
in Fig. 2.1a. The specimen arm is moved towards the knife by means of a
vertical pivot arm whose lower end rests on a lead screw and is advanced
automatically in regular increments from 25 nm to 500 nm between each
cutting stroke (Fig. 2.8). One full turn of the hand wheel moves the specimen
arm through a complete cycle and the only control of the cutting speed is the
rate at which the wheel is turned. The knife stage enables the knife to be
tilted, rotated and moved laterally and it can be advanced towards the specimen arm by coarse or micrometer adjustments. A certain degree of orientation

of the specimen is possible by means of a universal or ball-joint connec-
tion between the specimen block holder and the specimen arm. The binocular microscope is carried on a pivot arm attached to the base plate on
which the ultramicrotome stands and a fluorescent lamp is mounted on top
of the instrument.

The success of the MT1 led to the design in 1962 of the automatic MT2.
Several improvements were made in this model although the same principles
of construction are used. The main differences are that the arm movements
have been motorised giving continuously variable cutting speeds between
0.09 mm and 3.2 mm per second. The hand wheel is still incorporated for
manual operation and can be used to override the motor if desired. The
distance between the two pivots A and B (Fig. 2.8) can be adjusted on this
model giving a wider range of advance, from 1 nm to 4000 nm per stroke.
During the return stroke the specimen arm is retracted backwards 0.6 mm

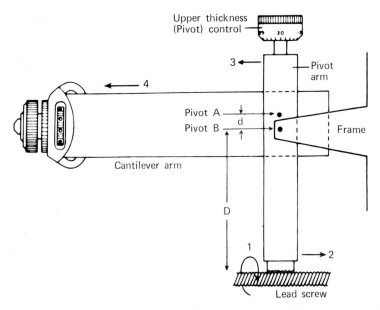

Fig. 2.8.   Diagram of the advance system of the Sorvall 'Porter-Blum' MT2 ultramicrotome (viewed from the side). The cantilever arm moves in a vertical plane about pivot B.
When the lead screw 1 is turned the bottom of the pivot arm moves in the direction 2
causing the upper end to move in direction 3. The cantilever arm, attached to the pivot
arm at pivot A, is in turn moved in direction 4. The magnitude of the movement is governed
by the ratio d/D and the amount the lead screw is rotated. The upper thickness control
varies the distance (d) between the pivots and so extends the range of advance. This control
is not present in the MT1 model. (By courtesy of Du Pont Company, Sorvall Operations).

from the knife edge (Fig. 2.2). The knife movements are similar to those of the MT1 but the specimen holder has an additional adjustment so that the block can be moved through a calibrated arc in a vertical direction. The viewing arrangements are almost identical to those of the MT1.

Fig. 2.9. The Sorvall 'Porter-Blum' MT2 ultramicrotome. (By courtesy of Du Pont Company, Sorvall Operations.)

The latest version, the MT2B, has two additional controls for setting the start and the length of the calibrated cutting stroke. When sectioning a small area the length of the cutting stroke can be set to the minimum so that the number of sections cut per minute is increased.

### 2.5 LKB Ultrotomes I and III (Figs. 2.10–2.12)

The first ultramicrotome manufactured by LKB was the Sjöstrand ultra-microtome (Sjöstrand 1953), and the experience gained from this led to the development by Hellström and Persson of a completely new instrument, the Ultrotome I (Hellström 1960). The specimen arm consists of a long hollow steel rod connected at one end to the foundation block by two flat phosphor-

Fig. 2.10.   Diagrammatic representation of the LKB Ultrotomes. (1) Knife stage; (2) Glass knife; (3) Knife holder; (4) Blade springs; (5) Specimen arm; (6) Specimen block; (7) Specimen holder; (8) Adjustable arc; (9) Specimen orientation head; (10) Moving-coil mechanism; (11) Foundation block; (12) Steel base-plate; (13) Damping cushions; (14) 'Macro-feed' knob; (15) 'Micro-feed' knob; (16) Collecting trough; (17) Heating coils; (18) Blower; (19) Retraction magnet; (20) Micro-feed beam. (By courtesy of LKB Produkter AB.)

bronze blade springs, in such a way that it is free to move in a vertical plane but not horizontally. A moving coil motor is connected to the specimen arm by means of a flexible terylene cord and lifts the arm to the top of its stroke. The cutting force is provided by the action of gravity but a range of cutting speeds between 1 mm and 20 mm per second is obtainable by varying the rate at which the coil motor is released. The specimen arm is then raised through the same path without any deflection. To achieve the desired clearance between the specimen block and the knife on the return stroke, the knife is retracted by applying an electromagnetic force to the platform carrying the knife stage (Fig. 2.3). The natural elasticity of the metal platform allows it to be bent downwards causing the knife to be moved backwards approximately 30 μm from the cutting point. When the specimen arm has cleared the knife edge on its return stroke, the magnetic action is released and the platform and the knife return to their original positions. This retraction takes place automatically between each stroke. The ultra-fine advance is achieved by the fine control of a heating coil which causes the specimen arm to expand towards the knife in regulated amounts from 5 nm to 100 nm between each stroke. This simple design of the specimen arm allows for a robust construction so that even with rough handling, damage influencing the advance cannot occur. To keep the expansion of the specimen arm within a linear range the arm can be quickly cooled after prolonged use of the thermal advance by drawing air through it with the aid of a fan blower. The coarse and fine advances of the knife towards the specimen arm are made by two independent mechanical controls. The specimen arm can be operated automatically, in single strokes, or by a remote manual control. A separate control unit minimises the risk of transmitting vibrations by hand when adjustments are made during sectioning. Rotation, tilt and lateral movements of the knife can be made and a large degree of orientation of the specimen block is possible (Fig. 2.11). The microscope stand carries a binocular microscope and a dual light system mounted on two arms both of which are moveable in an arc around the cutting point. A focused light source is provided for the critical examination of the knife edge and a fluorescent lamp for use during sectioning. The knife height is set by alignment to a reference peg and the specimen arm can be clamped in a mid-stroke position while the specimen block is aligned with the knife.

Due to improvements in techniques a need was felt for a more versatile instrument and this resulted in 1965 in the production of the Ultrotome III. The same principles of construction are used as in the Ultrotome I but there are certain refinements. An extended range of cutting speeds from 0.1 mm to

Fig. 2.11.   Possible adjustments to the knife and the specimen block in the LKB Ultrotomes. All movements centre around the cutting point. (By courtesy of LKB Produkter AB.)

20 mm per second, together with a longer cutting stroke, enable much larger
areas to be sectioned. The ultra-fine advance range has been extended to
130 nm and the thermal and mechanical stability has been increased to cope
with the long time intervals required for the very slow cutting speeds. Specimen blocks can be trimmed *in situ* on the specimen arm using glass knives

Fig. 2.12.    The LKB Ultrotome III. (By courtesy of LKB Produkter AB.)

and a planetary gear calibrated in 1 μm steps which is fitted to the knife advance. Incorporated with the focused light source is a device enabling the real angle of a glass knife to be measured at any point along its edge. An adjustable arm, called the section collector, is attached to the microscope and provides a holder for supporting the specimen grid during collection of the sections. Some of these items can be obtained as the UM II kit for use with the Ultrotome I.

An Ultrotome table is available, if required, for use with these instruments and can be filled with sand ballast to damp out certain transmitted vibrations. It also provides a convenient working surface and storage space for sectioning equipment. For more troublesome vibrations, in the frequency range 5–350 Hz, insulation can be provided by the LKB Ultrostable. An inner vibration free table is suspended in an outer frame by large coil springs. Ballast weights are added according to the weight of the instrument, to stabilize the system.

## 2.6   Reichert OMU2 and OMU3 ultramicrotomes (Figs. 2.13–2.15)

The first Reichert ultramicrotome, the OMU1 was designed by Sitte (1953) and had a thermal advance system with water cooling. This was superceded in 1964 by a new instrument the OMU2. The microtome is mounted on its own table with the controls set in the table top. One end of the specimen arm is connected to the feed system and mounted on ball bearings to allow movement in all directions. The arm movements are controlled by a motor which is supported under the table, remote from the instrument to avoid the risk of vibrations. The arm follows a guided path on the downward cutting stroke and is displaced sideways during the return stroke (Fig. 2.1b). The cutting speed can be varied between 0.5 mm and 10 mm per second and a separate control varies the speed of the return stroke, although neither are calibrated. A hand wheel supplies a manual control which can also be used to override the motor. The thermal advance is produced by a 25 W lamp heating a small block of metal of low thermal capacity which moves the specimen arm towards the knife in regulated amounts from 0 to 1000 nm per stroke as it expands. Expansion ceases when the arm movement is stopped and the arm is automatically retracted a small distance to prevent contact with the knife on recommencing sectioning. The knife stage has similar adjustments to those instruments already described, enabling the knife to be moved in several directions. The specimen holder can be rotated and moved through an arc in a vertical direction for alignment of the specimen block

Fig. 2.13. The Reichert OMU2 ultramicrotome. (By courtesy of C. Reichert AG.)

with the knife edge. An arm carrying a binocular microscope and a fixed fluorescent lamp is mounted on top of the instrument and can be swung to the side when mounting the knife and specimen block.

In 1970 a further instrument, the OMU3, was introduced, similar in construction to the previous model but with more refined movements of both the specimen holder and the knife stage. The table is much more streamlined and it has a useful set of drawers on one side for storage of accessories. Two other features are the trough filler for adjusting the level of the trough liquid and a spot light positioned below the knife edge to enable a more accurate approach of the knife towards the specimen to be made.

Fig. 2.14.   The Reichert OMU3 ultramicrotome. (By courtesy of C. Reichert AG.)

## 2.7   Tesla BS 490 ultramicrotome (Fig. 2.16)

This instrument is also mounted on its own table with the controls set into the table top. The specimen arm is motor driven and follows an elliptical path during its return stroke (Fig. 2.1b). By varying the speed of the motor, cutting speeds between 0.5 mm and 4 mm per second can be obtained. The instrument operates on the thermal advance principle with a range of advance of 5 nm to 170 nm per stroke and incorporates a memory device. This enables the expansion reached at the end of a sectioning period to be maintained

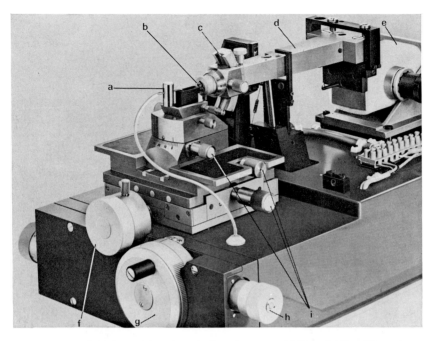

Fig. 2.15. Detail of the Reichert OMU3 ultramicrotome. (a) Trough filler; (b) specimen block; (c) adjustable arc; (d) specimen arm; (e) expansion block; (f) fine advance; (g) manual operation control; (h) trough filler control; (i) knife controls. (By courtesy of C. Reichert AG.)

until sectioning recommences, so that the distance between the knife and the specimen arm remains constant. The knife stage has the usual adjustments enabling the knife to be moved in several directions and the specimen block can be moved through an arc and rotated in a vertical plane. A binocular microscope and a fixed fluorescent lamp are mounted on an arm on the top of the instrument.

Only the main features of each of these ultramicrotomes have been described here; more detailed information can be obtained by contacting the manufacturers or their agents (see list of suppliers in Appendix). All these ultramicrotomes are capable of cutting sections thin enough for the electron microscope; it is in their mode of operation, versatility, reproducibility and of course price, in which they differ. Consideration should be given to what is required from such an instrument, both now and in the future before making a choice. It is worthwhile asking more experienced workers in the

Fig. 2.16.   The Tesla BS 490 ultramicrotome. (By courtesy of EXAL.)

field of ultramicrotomy for their advice but this should always be coupled, if possible, with a personal trial of the instrument.

REFERENCES

Fullam, E. F. and A. E. Gessler (1946), A high speed microtome for the electron microscope, Rev. scient. Instrum. *17*, 23.
Gettner, M. E. and L. Ornstein (1956), Microtomy, in: Physical techniques in biological
  research, Vol. 3, G. Oster and A. W. Pollister, eds. (Academic Press, New York), p. 627.

Hellström, B. (1960), 'The Ultrotome ultramicrotome' – basic principles and summarized description of construction, Science Tools (LKB) *7*, 10.

Huxley, A. F. (1957), An ultramicrotome, J. Physiol. *137*, 73.

Latta, H. and J. F. Hartmann (1950), Use of a glass edge in thin sectioning for electron microscopy, Proc. Soc. exp. Biol. Med. *74*, 436.

O'Brien, H. C. and G. M. McKinley (1943), New microtome and sectioning method for electron microscopy, Science, N.Y. *98*, 455.

Porter, K. R. (1964), Ultramicrotomy, in: Modern developments in electron microscopy, B. M. Siegel, ed. (Academic Press, New York), p. 121.

Porter, K. R. and J. Blum (1953), A study in microtomy for electron microscopy, Anat. Rec. *117*, 685.

Sitte, V. (1953), Ein einfaches Ultramikrotom für hochauflösende elektronenmikroskopische Untersuchungen, Mikroskopie *10*, 365.

Sjöstrand, F. S. (1953), A new microtome for ultrathin sectioning for high resolution electron microscopy, Experientia *9*, 114.

Von Ardenne, M. (1939), Die Keilschnittmethode, ein Weg zur Herstellung von Mikrotomschnitten mit weniger als $10^{-3}$ mm. Stärke für elektronenmikroskopische Zwecke, Z. Wissensch. Mikroskop. *56*, 8.

# Chapter 3

# Knives

## 3.1 The cutting edge

The two most important items in ultramicrotomy are the preparation of the of the specimen and the knife. Since the quality of these will limit the quality of the sections, care and attention to detail is most important for the beginner. As with most techniques, one learns with experience which short cuts can be taken but until that stage is reached any short cuts will normally have the opposite effect and increase the amount of work.

The first knives used in ultramicrotomy were carefully sharpened steel histology knives or razor blades but their edges lost their sharpness after cutting relatively few thin sections and the tedious process of sharpening had to be started again. In 1950 Latta and Hartmann discovered that a freshly broken piece of plate glass produced an edge suitable for cutting ultrathin sections and because of its relatively low price could be treated as disposable once it had become blunt. Glass is a supercooled liquid and the freshly broken edge is subject to molecular flow resulting in a rounding of the edge and therefore a loss of sharpness. The edge is also particularly susceptible to contamination from dust and can easily be damaged if touched. For these reasons glass knives should be prepared on the day of use and not stored for any length of time. Each section cut on the knife will reduce its keenness and with certain hard specimens the knife may only be capable of cutting three or four useful sections; some materials such as metals cannot be cut at all with a glass knife.

In the search for a more perfect cutting edge Fernández-Morán (1953, 1956) made a knife from a single crystal of diamond. The crystalline structure provides a very sharp, stable edge of molecular thickness and of unsurpassed

246

hardness, and is capable of cutting almost any material. The first commercial diamond knives did not always prove successful but techniques have improved and with several manufacturers now producing these knives (see Appendix) their popularity is increasing. The cost of a diamond knife is very high compared to that of a glass knife and is due to the exacting processes involved in its production. Prices vary according to the angle of the knife and the amount of useable edge.

Although a diamond knife edge is extremely hard it can very easily be damaged by misuse. It is therefore inadvisable for anyone to begin ultramicrotomy with a diamond knife without having had previous experience using a glass knife. Diamond knives vary considerably in performance and it is well worthwhile paying a little more to have one selected for a particular type of specimen. A sample specimen block should be sent together with the order to the manufacturer who, having selected a knife, will send the results of sectioning as a guarantee of the knife's suitability. Although several experiments have been made no other cutting edges have so far been found suitable for ultramicrotomy.

## 3.2  Preparing the glass to make knives

A sheet of glass is first broken into suitably sized strips from which the knives will be produced. Several different types of glass have been tried but good quality plate glass, from 5 mm to 6.5 mm thick, is generally adequate. It is however preferable to use it in the unpolished state as the polishing process introduces stresses into the surface of the glass. Plate glass is either 'drawn' through rollers or made by the float process. In the first method the glass will have stresses parallel to the direction in which it has been 'drawn'. More reproducible results will be obtained if the strips of glass are formed parallel to these stresses. Unfortunately it is difficult to know from examining the glass in which direction these stresses lie, unless it has been marked by the manufacturer. Float glass is made by floating the molten glass onto a liquid surface and is of a more uniform quality. It is usual nowadays to buy the glass already prepared in suitably sized strips which have been cut by a local glass merchant or obtained from a company preparing them especially for ultramicrotomy (see Appendix). The latter may be more expensive but will generally be of a better quality.

If it is necessary to prepare the strips in the laboratory a sheet of glass approximately 40 cm long and of a width that is a multiple of the final strip width, normally 25 mm to 38 mm, will be required. A straight score, using

Fig. 3.1.   A glass sheet is scored in straight lines from one side to the other. The first break is made in the centre and then each resultant piece is broken in half.

a scoring wheel and a straight edge, is made down the centre of the glass sheet dividing it into two equal pieces (Fig. 3.1). A scoring wheel is used in preference to a diamond cutter as the wheel forms a cleft forcing the glass apart so that less force is required to break it. The score is then placed directly over a pressure point and a downwards pressure applied on either side of the mark.

The breaking pressure can be applied by hand and a small wooden stick used to provide the centre pressure point (Fig. 3.4); or by using a pair of pliers with one or three raised pressure points, as first described by Weiner

Fig. 3.2. A pair of pliers modified with three strips of adhesive tape to provide three point contact. It is important that the two outer strips are equidistant from the central one.

(1959). One pressure point is made by taping a small wooden stick (a piece of matchstick or tooth pick) to the centre of one of the jaws of the pliers. The other jaw should be covered with adhesive tape to prevent the glass slipping. Three short strips of tape can be used to provide a three point contact (Fig. 3.2) or specially shaped pliers are available commercially (see Appendix). The centre raised point is placed directly under the score and a steady even pressure is applied until the break begins, the pressure is then maintained so that the break continues as slowly and evenly as possible. As the break nears the opposite side, the resistance offered by the unbroken glass decreases rapidly and the speed of the break will accelerate. On examination of the broken edges it will be seen that there are deep stress marks at the two extremities where the break has been fastest (Fig. 3.3b). This process is then continued, thus dividing each resultant piece of glass in half until the required strip size is reached. If the strips are broken starting from one side of the sheet of glass it will be difficult to get an even balance during breaking and the resulting twisting effect will produce pronounced stress lines on the broken surfaces. When the breaking force is very large the glass disperses the energy by increasing the amount of broken surface. This can be seen as needle-like projections on the surface of the glass, called hackle or needle marks (Fig. 3.3b).

The use of the free break method is still sometimes advocated. A short score 2 cm to 3 cm long is made near one edge of the glass sheet and the breaking pressure applied by one of the methods already described. The break will open up along the score line but then continue unguided across to the opposite side of the glass sheet (Fig. 3.4). Although the freshly broken edge will be free from score marks the disadvantage of this method is that the break rarely follows a straight line because of the difficulty of obtaining an even distribution of pressure. Knives made from strips produced by the free

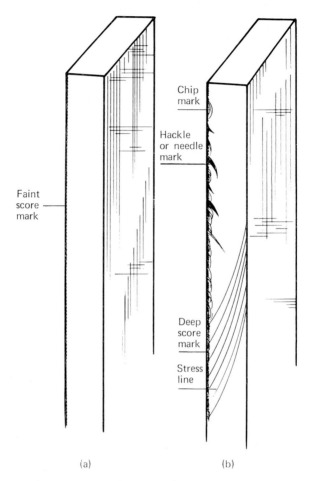

Chip
mark

Hackle
or needle
mark

Faint
score
mark

Deep
score
mark

Stress
line

(a)                                    (b)

Fig. 3.3.   (a) The appearance of a good glass strip; the only disturbance is a faint score mark on the left-hand side. (b) Chip marks resulting from too deep a score, and stress lines due to a fast or uneven break are present. If the breaking force has been very large the glass has an irregular surface, seen as hackle or needle marks.

break method will vary considerably in shape and angle so that any attempt to reproduce sectioning conditions later will be impossible.

### 3.3   The production of glass knives

The technique of obtaining a good knife involves making an accurately positioned score and applying the minimum amount of pressure in the

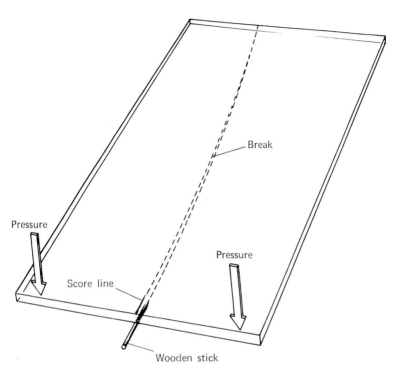

Fig. 3.4. The free break method of dividing a sheet of glass into strips. The break follows the short score line and then continues unguided across to the opposite side. The method of breaking shown involves the use of a wooden stick as the centre pressure point with hand pressure applied at each side of it.

correct place in order to give a slow controlled break. Breaking knives by hand is a skill not easily acquired by everyone and much time can be wasted by a beginner on such a task. The LKB KnifeMaker performs the functions of both scoring and breaking mechanically and is regarded today by many laboratories as an essential piece of ultramicrotomy equipment. Many methods for the production of glass knives have been described in the literature but the following is a method that can be used either by hand or adapted for use with a machine.

A strip of glass, free from dirt, grease and excessive stress marks is placed scored side downwards on a flat surface. If necessary the glass should be washed with soapy water and dried with a clean cloth. Care should now be taken in handling the glass to prevent finger marks being made on the two narrow surfaces (Fig. 3.5a). Using a scoring wheel and a straight edge, squares are then marked off along the strip. The score lines should start close to the one edge of the glass stiip but care should be taken to avoid chipping it. The end of the score may be taken right to the other side of the glass as this side will not normally be used to produce a knife. The glass is then firmly held while pressure is applied directly under the score until the glass breaks. This is most easily done in the hand method by using a pair of

Fig. 3.5.   The stages in the production of glass knives. (a) The glass should be clean, particularly at the surfaces (S). (b) The glass is divided into squares; the score starts about 1 mm from the edge of the glass. (c) The centre raised point of the pliers is placed directly under the score line. (d) The order in which squares should be broken. (e) The positioning of the diagonal score.

pliers with three point contacts as previously described (Fig. 3.5c). As with the production of glass strips straighter edges will be produced if the glass is always broken into two equal sized pieces (Fig. 3.5d). A score line is then made across the diagonal of the glass square. As it is almost impossible to bisect the corner evenly it is usual to deliberately offset the score to one side of the corner. The side chosen should be the one having the upper edge of the glass free from score lines or excessive stress marks. The score line should be placed as close to the edge of the glass as is practical (1 mm to 2 mm), and approximately 0.5 mm to one side of the corner (Fig. 3.5e). Pressure is applied as before to cause the break which will follow the score line as far as it goes and then will take the shortest route and curve towards the edge (Fig. 3.6). This results in one flat topped counter piece and one cutting edge with the real knife angle being greater than the scoring angle.

Several devices for applying a more controlled pressure to cause the break have been described in the literature and André (1962) describes an apparatus

Fig. 3.6. Enlargement of the corner of a glass square showing how the knife angle is greater than the scored angle because the free break curves away from the corner.

Fig. 3.7.   Shape of knife required for the Cambridge Huxley Mark 1 and Porter-Blum MT1 ultramicrotomes. If the knives are to be made by machine, a square of glass 38 mm wide is used to produce a triangular knife and the base is then shortened by breaking away a piece by hand.

for producing a more accurately positioned score. Van Iterson used the jaws of a machine vice to cause the break and a slightly improved method is described by Cosslett (1962). Fahrenbach (1963) describes a device called the 'Nutcracker' where the glass is squeezed between two hinged brass plates with pressure points. A rather more sophisticated instrument based on a similar idea was patented by Sutton (1969). In 1964 LKB introduced the KnifeMaker which is described in more detail in §3.4. A Japanese instrument, the Sunkay Messer, is also available commercially. In this instrument a square of glass of side 100 mm is placed in position on two rubber pads on the base of the instrument. The main bracket with two raised pins is lowered onto the glass and clamped in position and a score is made using a diamond cutter through a slot in the bracket on earlier instruments, or with the integral scoring wheel on the latest model. Rotation of a pressure knob causes a fulcrum to be raised under the score until the glass is broken. The large square is first broken into two rectangles which are then each broken in half. The resultant small squares are then positioned in turn so that the score is made across the diagonal. The results are dependent on the accuracy with which the glass is positioned over the fulcrum. An adjustable slot for holding one corner of the square and guide lines are provided for this purpose. Another knife making device, the Taab-Pyper, consists of a clamp for holding a square of glass in position whilst a short score is made with a hand-held scoring tool. The glass is then positioned in a vice-like clamp and gradual pressure applied to break it.

Both the Cambridge Huxley Mark I and the early Porter Blum MTI ultramicrotomes require knives of a different shape (Fig. 3.7). A jig is normally supplied with the instruments to provide an accurate guide for scoring when using hand breaking methods. The principle used in machine breaking, however, results in a knife in the shape of an isosceles triangle and consequently a square with a side of 38 mm is used to produce a larger triangular knife, the base of which is then shortened by breaking away a small piece by hand (Fig. 3.7). Alternatively the knife holder of the ultramicrotome can be enlarged to accept the large triangular shape (Biddlecome et al. 1971), or an adapter fitted to the knife holder which will then accept the smaller triangular knives (25 mm) described earlier (Griffin 1968; Marshall and Sheen 1970).

**3.4    The LKB KnifeMaker (Figs. 3.8 and 3.9)**

This instrument combines the functions of scoring and breaking the glass into squares or rhombi and then into knives of various angles. Two spherical headed pins on the clamping head press the glass onto two similar pins located in the KnifeMaker table and a spring loaded tungsten carbide scoring wheel is then drawn across the upper surface of the glass directly over the lower pins. Once the score has been made these lower pins are slowly raised by turning the breaking knob until the glass breaks. The scoring pressure can be easily adjusted and the position of the score varied by fine movements of the glass in relation to the scoring tool.

If strips of glass 40 cm long are used they should first be broken in half to provide a more suitable length. The clean strip is placed over the lower pins with one edge resting against the angle setting plate and the right hand end of the strip level with the spot marked on the KnifeMaker table. With the glass in position the clamping head is lowered onto it and providing no hand pressure is applied to the end of the strip the glass will automatically be levelled. It will be found that the clamping head is quite loose at this stage to allow for different thicknesses of glass but further downwards pressure on the clamping lever will pull the head sideways tightly against the clamping block. The shortest of the three scoring positions is selected and the scoring wheel pulled smoothly across the glass. Pressure on the scoring wheel should be adjusted to obtain a distinct score although it must not be so heavy as

Fig. 3.8. The LKB KnifeMaker. (a) Clamping lever; (b) arresting studs; (c) clamping head; (d) scoring tool; (e) angle setting plate; (f) breaking knob. (By courtesy of LKB Produkter AB.)

to produce splinters of glass. The breaking knob is then turned until the glass breaks; if it does not break it is because it has not been properly clamped. A square or rhomboid is then formed by placing one end of the strip against the arresting stud and one edge against the angle setting plate, and then scoring and breaking as above. Because of the greater control of the breaking pressure possible with this instrument, it is not necessary to break the glass into equal-sized pieces each time. It will however be seen that these broken surfaces are not completely flat but have a small 'lip'. This will not affect the knife as the cutting edge will not be formed on that side of the corner. Only strips of glass 25 mm or 38 mm wide should be used as the arresting pins are positioned to measure off only 25 mm or 38 mm pieces from the end of the strip. The angle setting plate can be positioned to provide a range of angles between 70° and 110°, but attention must be paid to the orientation of the glass (see instruction manual); for angles greater than 90° the scored edge of the strip should be placed uppermost.

The square or rhomboid is next orientated as in Fig. 3.5e and placed between the two sliding guides (Fig. 3.9). A longer scoring position is now selected for either the 25 mm or 38 mm wide glass and a score made across the diagonal. The front guide is set to bring the score 1 mm to 2 mm from the edge of the glass and the rear guide, which is spring loaded, adjusted so that the glass is under slight pressure. If the glass is not securely held in this way it may move out of position when the breaking pressure is applied resulting in irregularly shaped knives. Lateral adjustments of the guides move the glass in relation to the scoring line so that the score may be offset from the corner as already described. Breaking is accomplished as before by turning the breaking knob. The quality of the edge will be determined by the speed of the break and this can be judged by the sound it makes. A soft sound is the result of a slow break and will normally provide a good knife; a fast break produces a sharper crack and usually indicates that the scoring pressure needs increasing or the scoring wheel replacing. If the score has been correctly positioned two knife edges will be formed on opposite sides, 0.5 mm to 1 mm from each corner.

When using the damping pad (Fig. 3.9) as much as double the normal usable knife edge can be obtained on the knife at that side (see §3.5). The score is made first to ensure that it is correctly positioned and the rubber damping pad is then pushed onto the corner of the glass until a slight resistance is felt. This pad reduces the vibrations that are set up during breaking and so decreases the number of irregularities produced on the knife edge. If no improvement in the quality of the knife edge is seen a little more pressure should be applied to the damping pad next time. If too much pressure is applied or the knife edge is formed very close to the corner the damping pad will touch the freshly broken knife edge causing damage or contamination. Slight adustments may be needed to the various settings on the KnifeMaker in order to counteract variations in the glass but otherwise the results should remain reproducible.

### 3.5 Evaluation of the knife edge

Good sections can never be cut on a bad knife. Critical examination of the knife edge will reveal irregularities but not the degree of sharpness; this can only be judged by sectioning. Sharpness relates to the radius of curvature where the two facets of the knife meet at the cutting edge and is dependent

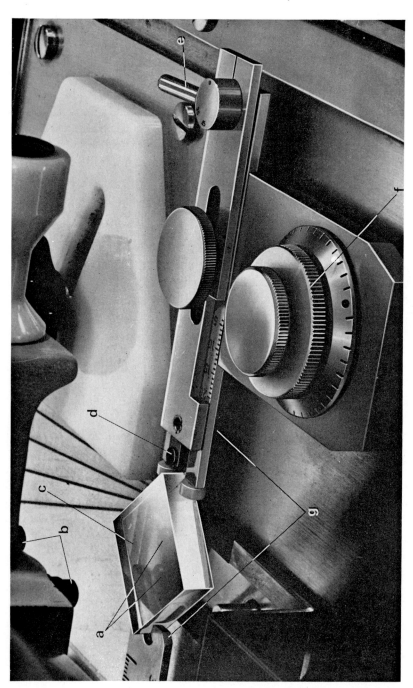

Fig. 3.9. Detail of the LKB KnifeMaker showing a glass square in position between the sliding guides. The rubber damping pad is not placed in contact with the corner of the glass square until after the score has been made. (a) Lower clamping pins; (b) upper clamping pins; (c) glass square; (d) rubber damping pad; (e) damping pad lever; (f) front lateral adjustment.; (g) sliding guides.

(By courtesy of LKB Produkter AB.)

on the molecular arrangement in the glass. Where there is a defect in the cutting edge the sharpness will not be the same and the section will be marked by a scratch. As a glass knife is rarely completely free from such defects it is important to know what part of the knife is usable, if any. These defects can only be seen clearly with a light focused on the cutting edge when it is dry. It is preferable to examine the knife immediately prior to sectioning so that the useful region may be determined. Good knives often get damaged by careless handling, particularly when attaching the trough, yet this often passes unnoticed. With ultramicrotomes which are equipped with a focused light source it is possible to perform the examination of the knife edge with the knife mounted on the instrument; alternatively a suitable binocular microscope with incident light illumination can be used (Sheldon 1957; Squier and Randall 1965). A beam of light is focused onto the knife edge so that it appears as a narrow band of light showing against a dark background. Irregularities will be shown quite clearly since they scatter the light. Only the area free from imperfections, when viewed at a magnification of about

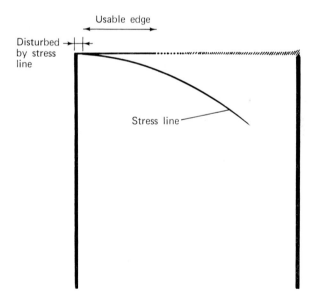

Fig. 3.10. Diagrammatic representation of the edge of a glass knife. When viewed in dark-field illumination the edge will appear white against a black background. The stress line indicates the side where the break started and the irregularities increase in magnitude towards the opposite side.

30 ×, should be used for ultrathin sectioning. This region is found to be on the side where the break began and the breaking speed of the glass was at its slowest; the magnitude of the defects will be seen to increase in the region where the break accelerated (Fig. 3.10). A prominent stress line will also be seen originating from the start of the break and prevents the extreme corner being used for sectioning. The amount of usable edge will depend on the angle of the knife and the method of breaking. Areas showing small defects will produce some scratches on the sections but may be used for 'thick' sections or for trimming the specimen block. The straightness of the edge does not affect its sharpness although it will be found to be more difficult to align the specimen block with a curved knife edge. Knives should be made just before use; if stored they should be kept under cover to prevent contamination by dust etc. The edges are also affected by moisture, so if they have to be kept for any length of time a desiccator should be used, but storage is not recommended.

## 3.6   The knife angle

As discussed earlier the knife angle will have an effect on the quality of the sections. The actual knife angle is determined to a certain degree by the position of the score when making the knife but the amount of curvature (Fig. 3.6) cannot be controlled easily. Measurement of the real knife angle can be made by splitting a beam of light on the knife edge and measuring the angle of separation. A simpler method is used on the LKB Ultrotome III on which an attachment to the lighting arrangement allows accurate measurements of the knife angle to be made very easily. The knife and the binocular microscope are both placed in the vertical position (Fig. 3.11) and a narrow beam of light is focused onto the sectioning facet of the knife. Then the lamp is moved through an arc above the knife edge until the centre of the beam of light is reflected from that part of the knife edge which is to be measured (Fig. 3.12). The position of the lamp arm indicates the knife angle which can be read directly on a scale. It is essential that the knife edge be placed at the correct height as indicated by the reference peg, to ensure that it lies at the centre of the arc described by the lamp. Compensation must be made to the reading if either the microscope or the knife are tilted from the vertical position. It may be of interest to note that the actual angle will generally be about 10° greater than the scored angle from knives made on the Knife-Maker and with hand broken knives the actual angle may be as much as 25° greater.

Fig. 3.11. Diagram illustrating the method used for the measurement of the real knife angle in the LKB Ultrotome III. The focused lamp is positioned to give the correct reflection from the knife edge (Fig. 3.12). If the microscope is in the vertical position (zero on the outer scale) and the clearance angle is zero, the position of the focused light (as indicated on the inner scale) corresponds to the real knife angle. If the microscope is tilted X° towards the lamp, X° must be subtracted from the lamp reading. Similarly if the clearance angle is Y° this figure must also be subtracted from the lamp reading. In the unlikely event that the microscope or the knife are tilted in the opposite directions their positions should be added to that of the lamp reading.

Area being measured

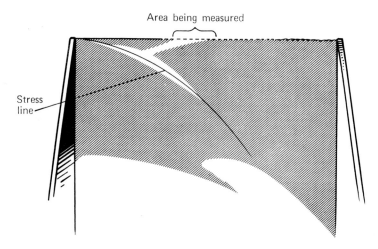

Stress
line

Fig. 3.12.   A beam of light positioned on part of a glass knife edge for measurement of the
actual knife angle. If the beam of light is not parallel to the knife edge the angle of the
knife will vary along the edge. The light reflected from the stress line is ignored.


## 3.7   Troughs

To enable the sections to be floated onto a liquid surface a leak proof trough
is mounted at the knife edge. These troughs are usually made of pre-shaped
metal or an adhesive tape. A metal trough clips round the glass knife and is
then sealed with wax to make it water tight. Dental wax, which can be obtained from any dental supplier, is preferable to ordinary paraffin wax as it
is cleaner and less brittle. A small piece of the wax is placed inside the trough
which is then gently heated causing the wax to melt and flow into all the
crevices by capillary action (Fig. 3.13). After use the troughs can be soaked
in acetone or heated to remove all the old wax.

Tape troughs have the advantage of being disposable but will normally
provide a smaller, although adequate, liquid surface. Care should be taken
when selecting a tape as some adhesives may be affected by the trough
liquid and form a film on the liquid surface. Scotch electrical tape No. 33
and polyester film tape No. 850 silver (see Appendix) have proved very satisfactory for making troughs. The tape should not be so wide as to be held
by the knife clamp when the knife is mounted on the ultramicrotome, since
the clamp may break the seal as it is tightened. One end of a short piece of
tape is cut to a point and then pressed firmly onto one side of the knife.
The top edge of the tape should form an angle of approximately 90° with the

clearance facet of the knife and not protrude beyond it or part of the specimen block may touch the tape as it passes the knife and break the seal. The other end of the tape can be cut to a point and attached in a similar way to the other side of the knife or, if difficulty is encountered in judging the correct length, the excess tape may be trimmed away with a razor blade (Fig. 3.14). Care should be taken to avoid touching the knife edge with the razor blade. It is not advisable to cut tape in this way on the side of the knife which is closest to the usable part of the knife edge. Molten dental wax is applied on the outside of the trough with a spatula or small brush to form the seal as it is not possible to heat the tape in the same manner as the metal troughs. Some workers advocate the use of nail lacquer as a sealant; this is easy to apply but takes a little longer to dry (Fernández-Diez and Gonzáles-Angulo 1970).

A very simple 'trough' can be made by using a mark from a grease pencil, or a blob of wax, to retain a drop of water at the knife edge. This provides a rather small, convex surface making it difficult to view the sections; it is rarely used now for ultrathin sections but is a quick method for collecting thick sections. Westfall (1961) cemented a wedge of bakelite to the sectioning facet of a knife just below the knife edge. This provided a level surface to support an elongated drop of water with an almost flat surface. Gemmell and Henrikson (1970) describe a large trough for collecting sections directly

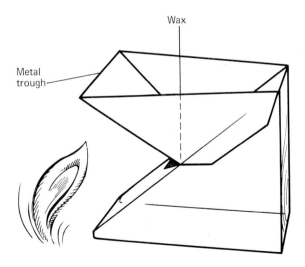

Fig. 3.13.   A glass knife fitted with a metal trough. The back of the trough is gently warmed to melt the wax inside.

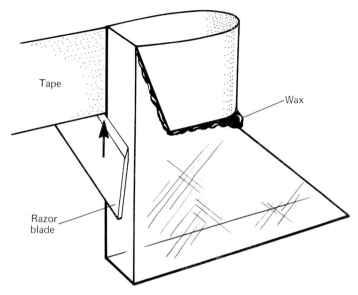

Fig. 3.14.   A tape trough being applied to a glass knife. The razor blade is only used to cut the tape on the side furthest from the usable region of the knife edge.

onto a microscope slide for autoradiography. This obviates the need for additional handling in transferring the sections and could be used for the direct mounting of thick sections.

## 3.8   Diamond knives

Diamond knives are used to cut specimens that are too hard to be cut with glass knives, such as calcified tissue and metals. They also have the great advantage that the cutting edge remains sharp for a considerable period of time. This makes them ideal for use when very long series of sections are required and it is inconvenient to keep changing knives, with the consequent re-alignment that would be necessary if glass knives were used. A good diamond knife will also be capable of cutting thinner sections, 10 nm–20 nm thick, which will be needed if very high resolution is required. Although a diamond knife edge is extremely hard and durable it is also very brittle and can easily be damaged by misuse. The edge is obtained by the cleavage of certain industrial diamonds along suitable natural planes, followed by grinding and polishing with ultrafine abrasives. The diamond is sintered into a metal shaft for handling purposes and this shaft is then fixed in a diamond knife holder incorporating a trough. Some manufacturers cement the metal shaft into a trough and others supply it separately. As certain ultramicrotomes require special holders this should be taken into account when ordering a diamond knife. Each knife should have its own holder and should not normally be removed from it as continual handling will increase the risk of damage to the knife. The included angle of the diamond knife varies between 40° and 60°, but the exact angle is difficult to measure. The smaller angles are used for biological work and the larger angles, which provide a stronger knife, are recommended for sectioning metals and other hard materials. The correct clearance angle for each knife is determined by test sectioning or from the manufacturer's instructions as the position of the knife in its holder will be different for each knife. A block of plain embedding material can be used for the test sectioning which is started using a low clearance angle; the angle is then increased in 1° or 2° steps until the optimum results are obtained. This angle can be used as a starting point for other specimen blocks. Small variations of the clearance angle may be necessary from one specimen block to another, and it should be noted that the angle of the diamond knife may vary 1° to 2° along its length. Quite often, when using a diamond knife at a low clearance angle, the specimen block may be seen to rub on the clearance facet of the knife. The reason for this is that this facet of the knife is not straight but has irregularities produced by the grinding process during manufacture. When this rubbing occurs the clearance angle should be increased until the specimen block can pass without interruption.

Care should be taken to avoid cutting sections thicker than 1 μm on a diamond knife, particularly of hard materials, as this will quickly damage the edge. This is especially important when fast sectioning speeds are used or the cutting force is produced by a motor. Diamond knives should never be used for trimming unless they have been discarded from normal use. Sections should always be removed from the knife edge so that they cannot dry onto the sectioning facet of the knife as the trough liquid evaporates, because they are then very difficult to remove. During sectioning small particles of the embedding material may adhere to the knife edge and these will cause scratches on the sections (see §7.6), if they are allowed to remain. The edge should be cleaned with a stick of soft wood, balsa or pith, which has been dipped into acetone or alcohol. One end of the cleaning tool is cut flat and then drawn across the whole knife edge with a gentle but steady pressure so that the knife edge just cuts into the cleaning stick. This wiping movement must only be made parallel to the edge (Fig. 3.15); *never* obliquely or at right

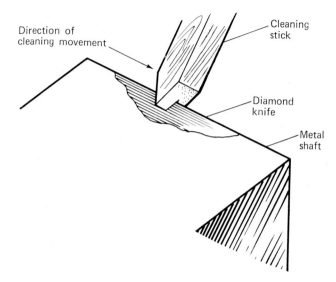

Fig. 3.15.   Cleaning a diamond knife with a soft wood stick. All movements must be made parallel to the knife edge.

angles to it. If the edge is very dirty a harder material such as a stick of plexiglass may be used in the same manner.

The life of the cutting edge will be dependent on the hardness of the material being sectioned and on the care the knife receives; it can be anything from a few weeks to several years. The edge should never be touched by anything other than the specimen or the cleaning tool and to ensure that the correct care is given to a knife each operator should have his own knife, if possible. Re-sharpening has to be done by the manufacturer although it may not always prove successful; it is also expensive and can involve a period of several months. The process of re-sharpening causes an increase in the angle of the knife and it is sometimes found that the knife is no longer suitable for sectioning the same type of specimen block. One of the difficulties that is encountered when using a diamond knife is to know when the knife needs re-sharpening. If the sections obtained are not satisfactory, the fault may lie with the knife or the specimen. If it is possible to use a glass knife this should be tried and if the results prove much better it will indicate that the problem lies with the knife. If the sections are badly compressed or creased when using an old diamond knife, it will indicate that the edge is in need of re-sharpening. If the problem occurs with a new diamond knife the reason is probably that the knife angle is not suitable for that particular specimen.

Another symptom of the incorrect angle being used is seen as narrow bands of varying thickness within a section only visible in the electron microscope, so-called 'chatter' (§7.9). If this cannot be avoided by varying the clearance angle the only solution is to select another knife. Ideally a selection of knives with different angles should be available.

After use unwanted sections should be removed immediately while the knife is still wet. The knife and the trough are then carefully rinsed with distilled water and allowed to dry. If debris remains, acetone or alcohol can be used to clean the trough and the knife edge cleaned as described above. The use of other cleaning compounds has been described in Du Pont's Ultramicrotomy Newsletter Nos. 1, 2 and 3 (see Appendix for address), but it is important to remember that the knife and trough should both be thoroughly rinsed again with water before drying. Acid should never be used to clean a diamond knife. The clean knife should then be left in its holder and stored in its box or in a safe dust-free place.

REFERENCES

André, J. (1962), Présentation d'un appareil destiné à préparer les couteaux de verre, J. Ultrastruct. Res. *6*, 437.

Biddlecombe, W. H., K. J. Ballard and H. Y. Elder (1971), Adaption of the standard Cambridge Huxley ultramicrotome knife clamp to accept triangular machine-made glass knives, J. Microscopy *93*, 163.

Cosslett, A. (1962), Preparation of glass knives for thin-section cutting, Jl R. microsc. Soc. *80*, 301.

Fahrenbach, W. H. (1963), A contribution to glass knife breaking, J. Cell Biol. *18*, 475.

Fernández-Diez, J. and A. Gonzáles-Angulo (1970), Sealing troughs to glass knives: advantages of fingernail lacquer (nail polish), Stain technol. *45*, 190.

Fernández-Morán, H. (1953), A diamond knife for ultrathin sectioning, Expl Cell Res. *5*, 255.

Fernández-Morán, H. (1956), Applications of a diamond knife for ultrathin sectioning to the study of the fine structure of biological tissues and metals, J. biophys. biochem. Cytol. *2*, suppl. p. 29.

Gemmell, R. T. and R. C. Henrikson (1970), A large trough for mounting thin sections on slides, Stain technol. *45*, 135.

Griffin, R. L. (1968), Machine-made glass knives in electron microscopy, J. med. Lab. Technol. *25*, 146.

Latta, H. and J. F. Hartmann (1950), Use of a glass edge in thin sectioning for electron microscopy, Proc. Soc. exp. Biol. Med. *74*, 436.

Marshall, J. and F. Sheen (1970), A modified knife holder for the Cambridge Huxley ultramicrotome, J. Microscopy *92*, 155.

Sheldon, H. (1957), A method for evaluating glass knives, J. biophys. biochem. Cytol. *3*, 621.

Squier, C. A. and M. Randall (1965), A simple adjustable holder for the inspection of glass knives, Jl R. microsc. Soc. *84*, 571.

Sutton, J. S. (1969), Producing improved glass knives for ultramicrotomy; a glass breaker featuring a linear fulcrum and a device for controlling fracturing velocity, Stain technol. *44*, 287.

Weiner, S. (1959), A new method of glass knife preparation for thin-section microtomy, J. biophys. biochem. Cytol. *5*, 175.

Westfall, J. A. (1961), Obtaining flat serial sections for electron microscopy, Stain technol. *36*, 36.

Chapter 4

# The correlation between light
# and electron microscopy

A great deal of time and effort may be saved by examining sections at lower magnifications in the light microscope before proceeding to the electron microscope. Specific areas may be located, orientation may be checked and uninteresting or poorly processed material may be discarded. It is far easier to cut thick sections than ultrathin sections and a lot of useful information can be obtained by examination of these sections under a phase-contrast microscope, or, after staining, under a normal light microscope. Methods for staining thick sections are described in detail by Lewis et al. (1974). Very often it is difficult for the beginner to relate what he sees at high magnification in the electron microscope to the material he sees with his own eyes. The light microscope is an intermediate stage and will help to give a better understanding of the material as a whole (De Bruijn and McGee-Russell 1966). Wells and La Cour (1971) describe a method of staining ultrathin sections mounted on a specimen support grid so that the same sections can be examined by both light and electron microscopy. Several models of electron microscope can be adjusted to provide a 'scanning range' with magnifications much lower than normal, approx $250 \times$ to $4500 \times$, which overlaps the range of the light microscope (Wisse et al. 1971). However ultrathin sections are more easily obtained if the area to be sectioned is small (see §5.1). Consequently when searching for specific regions within the specimen, it is advisable to cut thick survey sections of the whole area for light microscopy. When the region of interest has been located under the light microscope, the specimen block is trimmed down so that the block face only includes the required area.

**4.1   The production of survey sections**

Thick sections can be cut from trimmed (see Chapter 5), partially trimmed, or untrimmed specimen blocks. Early workers used modified ultramicrotomes (McGee-Russell et al. 1962), or standard histological microtomes adapted to accommodate glass knives and plastic embedded specimens (Behnke and Rostgaard 1963) for cutting survey sections. Later a simple microtome was built expressly for this purpose (Magill and Gunning 1969) and the LKB Pyramitome combines this function with that of specimen trimming. To be of real value the orientation of the specimen block should be maintained between the thick and ultrathin sections making the correlation more accurate. An attachment to the LKB Pyramitome, the Target Marker, enables the image of a stained thick section to be superimposed over the face of the specimen block. Fine detail is seen clearly in the section and the specimen block is then trimmed down precisely to the desired area (see § 5.3). All modern ultramicrotomes are capable of cutting thick sections at any stage before, during or after ultrathin sectioning. The thickness required for survey sections is normally between 0.5 µm and 2 µm. It is possible to section quite large areas at this thickness, the limitations being the width of the cutting edge and the length of the cutting stroke. Two micrometre thick sections can be cut quite easily using a dry knife but the thinner sections will be less creased if they are floated on a liquid surface. Troughs can be made in one of the ways already described.

**4.2   The collection of survey sections**

It will generally be found easier to handle one section at a time. If the section is floating on a liquid surface it is lifted out of the trough on a suitable support such as a piece of coverslip or a fine paint brush and then placed on a drop of water on a microscope slide. When using a brush it should be placed under the section and then carefully raised; the section will wrap itself around the brush but will unfold as soon as the section touches the water on the slide. The section may also be transferred by carrying it on a drop of liquid supported by a fine wire loop (Fig. 4.1a), or a thin foil annulus, as described by Marinozzi (1964). If the section has been cut on a dry knife and is of a reasonable size it can be picked up by grasping one corner of the section in a pair of fine tweezers. Smaller sections are more easily handled with the aid of a fine paint brush moistened with water. The brush is slid under the section (Fig. 4.1b) and treated in the same way as described above.

Fig. 4.1.   Collection of survey sections. (a) A section supported on a drop of water in a wire loop. (b) Lifting a section from a dry knife with the aid of a fine paint brush. (c) The section is placed on a drop of water on a glass microscope slide and gently warmed to remove any wrinkles.

If the brush is not moistened an electrostatic charge may cause the section to fly away.

Once the sections have been transferred the glass slide is gently warmed (Fig. 4.1c). This helps to stretch out any wrinkles that have been formed in the section and as the water evaporates the section becomes firmly attached to the slide and can then be stained (Lewis et al. 1974). The heat should not be too intense or air bubbles will be formed under the section and may become trapped as the section dries.

### REFERENCES

Behnke, O. and J. Rostgaard (1963), A device for mounting glass knives in the ordinary rotary microtome for sectioning plastic-embedded material, Stain technol. *38*, 299.
De Bruijn, W. C. and S. M. McGee-Russell (1966), Bridging a gap in pathology and histology, Jl R. microsc. Soc. *85*, 77.

Lewis, P. R., D. P. Knight and M. A. Williams (1974), Staining methods for thin sections, in: Practical methods in electron microscopy, A. M. Glauert, ed. (North-Holland, Amsterdam).

Magill, H. and B. Gunning (1969), A simple microtome capable of cutting sections of plastic-embedded material down to 1 μm thickness, J. Microscopy *89*, 217.

Marinozzi, V. (1964), Cytochimie ultrastructurale du nucléole-RNA et protéines intra-nucléolaires, J. Ultrastruct. Res. *10*, 433.

McGee-Russell, S. M., F. H. Roberts and J. K. Jacobs (1962), A modification to the Huxley ultramicrotome, Q. Jl Microsc. Sci. *103*, 405.

Wells, B. and L. F. La Cour (1971), A technique for studying one and the same section of a cell in sequence with the light and electron microscope, J. Microscopy *93*, 43.

Wisse, E., J. v. d. Meulen, J. M. van t'Noordende and J. J. Emeis (1971), Electron microscopy at low magnification. A histological application, Philips Bulletin EM 55.

# Chapter 5

# Trimming the specimen block

## 5.1 The correct size and shape of the specimen block face

Before sectioning the block, containing the embedded specimen, is trimmed down to a suitable size and shape. The size will normally be determined by the requirements of the experiment. Thick sections can be cut from areas 5 mm$^2$ or larger but when cutting ultrathin sections it is advisable to keep the block face as small as possible. This is particularly important when very thin sections or long series of sections are required. The only limit to the minimum area is the ease of handling the sections once they are cut. If as large an area as possible is required then the limiting factors will be the length of the cutting stroke of the ultramicrotome, the characteristics of the embedding material and the specimen, and the quality of the knife. It is possible with modern equipment to cut ultrathin sections as large as the specimen support grids but this is at the sacrifice of thinness, quality and quantity and cannot be considered a routine technique. The smaller the block face the easier it will be to cut ultrathin sections.

One of the most common reasons for difficulty in obtaining good sections is due to variations in the consistency of the specimen block. The area containing the specimen will often have a different hardness from the remainder of the block. Problems will be avoided if all the plain embedding material, which is of no interest, is cut away so that the specimen covers the whole area of the block face. Many specimen blocks that have been almost impossible to section have yielded good results after re-trimming in this way (Fig. 5.1). It is of course not always possible to do this but it will generally be found to be preferable to trim away as much of the embedding material as possible. A well embedded specimen will not give any trouble as it will be an even consistency throughout and a primary objective should be made to achieve this (Glauert 1974).

The shape of the block face can be varied according to the form of the tissue but if a straight ribbon of sections is to be obtained, the top and bottom edges of the block face must be parallel to one another. Also, if the edge of the block face that will hit the knife first (the bottom edge) is made slightly longer than the top edge, the previous section will be detached more cleanly from the knife edge. This shape also serves to identify the leading edge of the ribbon which is most important in keeping serial sections in sequence (Fig. 1.4). Good contact between the sections, providing a stable ribbon formation,

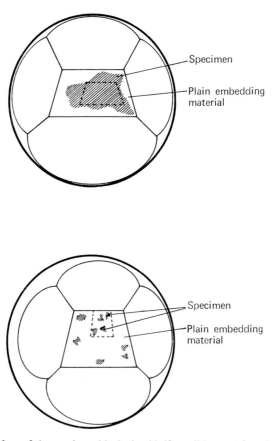

Fig. 5.1.   The face of the specimen block should, if possible, contain only the specimen. When the specimen is dispersed throughout the block an area containing the maximum amount of the specimen should be chosen. The broken lines indicate how these blocks should be re-trimmed if sectioning proves difficult.

is obtained if the width of the block face is approximately one and a half times its height. However, when difficulty is encountered in cutting longitudinal sections of specimens that have a certain structural orientation, e.g. nerve fibres, success is more likely to be obtained if the specimen is orientated with its long axis perpendicular to the knife edge (Fig. 5.2). This is because a more even strain is put onto a smaller region of the knife edge. The ribbon of sections should be kept fairly short when cutting in this way as there will be a tendency for the sections to separate from one another.

Trimming may be simplified by the correct choice of embedding mould. If small specimens are embedded in large moulds much more embedding material has to be trimmed away. There are several types of plastic mould available today which produce a pre-formed shape that requires a minimum of trimming (see Glauert 1974). When a specific orientation is required, flat embedding moulds can be used as there is then less chance of the specimen moving out of the desired position during polymerization of the embedding material and so making trimming more difficult. When using gelatin capsules as moulds it is easier to trim the specimen block if the gelatin is removed

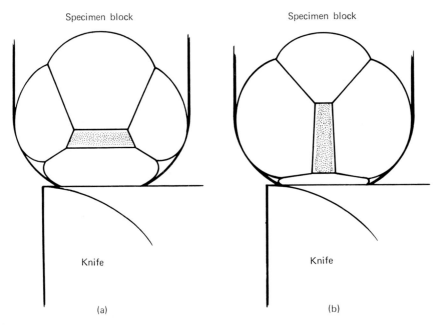

Fig. 5.2. A structurally orientated specimen will often vary in consistency and difficulty will be encountered if the long axis of the structure is parallel to the knife edge, as in (a). A more even strain is placed on the knife if the specimen block is orientated as in (b).

first by soaking the block in warm water to dissolve the capsule. Polymerized blocks can be easily removed from plastic moulds with the aid of a capsule press or splitter (see Appendix). The specimen block is then clamped in a suitably sized holder and placed in an orientation head, if one is available,

Fig. 5.3.   The specimen block should only protrude a few millimetres from its holder. An orientation head enables the specimen to be sectioned in the required plane.

so that the desired cutting plane can be chosen (Fig. 5.3). The end of the specimen block should not protrude more than a few millimetres from the holder to ensure adequate rigidity. Trimming can be done by hand, *in situ* on the ultramicrotome or with a trimming instrument.

## 5.2 Trimming by hand

It is possible on some ultramicrotomes to clamp the specimen block holder so that its axis is in a vertical plane and the specimen can then be viewed from above. Alternatively the holder can be held in a suitable clamp under a separate stereoscopic microscope. A simple but effective way is to replace the removable glass plate on the stage of the microscope with a disc shaped metal plate. In the centre of this is fixed a short cylindrical block which will accommodate the specimen holder. Fineran (1971) describes such a device for specimen holders from the LKB Ultrotomes and a similar method has been used for many years at the Strangeways Research Laboratory, Cambridge for the specimen holders of the Huxley ultramicrotomes.

The trimming tool, normally a single-edged razor blade, or a sharp blade mounted in a holder, is held firmly in one hand while the other hand supports the specimen block. A horizontal cut is first made across the top of the block to expose the specimen (Fig. 5.4a). The block may be hard to cut and great care should be taken to avoid the trimming blade slipping and cutting away too much material or one's fingers. Better control of the cutting action will be obtained if the hand holding the trimming blade presses against the other hand as the cut is made. Adhesive tape can be placed on the fingers holding the specimen block to protect them from accidental cuts if this is felt to be necessary. The trimming blade is then held at an angle of about 60° to the vertical and a cut made down one side of the specimen block (Fig. 5.4b). The complete holder is then rotated through 180° and the opposite side of the specimen block cut in a similar manner so that the two edges of the block face thus formed are as parallel as possible to one another (Fig. 5.4c). The other two sides are then cut in a similar way to produce a trapezoidal face to the specimen block (Fig. 5.4d and e). A smoother finish will be obtained if two or three slices are taken rather than one thick one and it may be found easier to cut a small step rather than continue the sloping cut to the edge of the block (Fig. 5.5). The angle of the sloping sides should be large enough to give adequate rigidity during sectioning without the section size increasing too rapidly as cutting progresses into the block.

The trimming blade should be discarded as soon as it becomes damaged, or deep scratches will be formed down the sides of the block producing a corrugated edge at the block face. This will prevent good adhesion between adjacent sections and also liquid from the trough may be drawn into these grooves by capillary action. It is worthwhile trying to judge the hardness of the block while trimming since this will give an indication of the conditions

Viewed from side                    Viewed from above

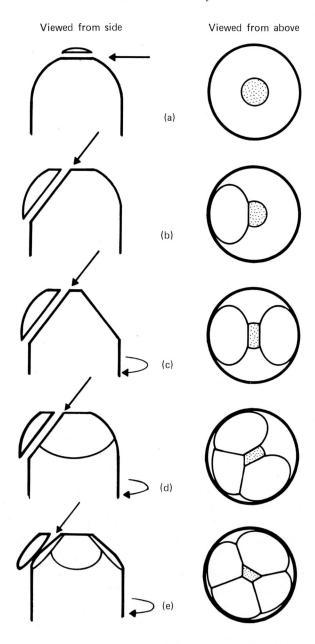

(a)

(b)

(c)

(d)

(e)

Fig. 5.4.   Stages in the trimming of a specimen block. The straight arrows indicate the
directions of cutting with the trimming blade.

Fig. 5.5.   Different methods of trimming a specimen block. (a) The sides are cut in a step. (b) The cuts are continued to the sides of the block.

needed later for sectioning. If the block is soft and can be trimmed easily it will often prove difficult to obtain ultrathin sections; a better block is one that is hard to cut with the trimming blade.

## 5.3   Mechanical trimming

Mechanical trimming will always give a more precisely shaped block face than any but the most skilful hand trimming and is strongly to be recommended for the beginner. Most of the methods described in the literature involve procedures which are too slow to be of general interest or leave a rough surface needing further smoothing. With certain ultramicrotomes the specimen block can be trimmed *in situ* by using a glass knife to cut thick sections from each of the sides in turn (Galey 1963). A specimen block is placed in a holder and mounted on the specimen arm. A glass knife is fixed in the knife holder and the knife stage adjusted so that the cutting edge is perpendicular to the specimen arm. Sections several micrometres thick are cut from the end of the block until the specimen is exposed. The knife stage is then rotated through approximately 30° to each side in turn (Fig. 5.6), and thick sections cut until the desired area is reached. This will produce the two parallel edges of the block face. That part of the specimen holder fitted to the specimen arm is now rotated to bring the other surfaces of the block into position to be trimmed. This method will produce very smooth surfaces and any scratches that occur on the sides of the block will be parallel to the edges of the block face and are unlikely to cause the difficulties described in the previous section (§5.2). Unless the specimen arm can be operated fairly quickly this technique may be too time consuming to be used routinely.

Fig. 5.6. Trimming a specimen block on the LKB Ultrotome III. (By courtesy of LKB Produkter AB.)

Care should be taken to avoid straining any part of the ultramicrotome by trimming in this way and if there is any doubt the manufacturer or his agent should be consulted.

Three trimming instruments are available commercially, all of which produce a well-shaped block in a short time. The Cambridge Block Trimmer (Fig. 5.7) uses an angular steel blade as the cutting tool and this is drawn across the sides of the block with each movement of a handle. The cutting blade is advanced by a feed-wheel between each stroke; the advance is not calibrated and is set by 'feeling' the depth of cut. The specimen block is held in a collet chuck which is rotated to pre-set positions presenting each side

Fig. 5.7.   The Cambridge Block Trimmer. (By courtesy of Kent Cambridge Medical.)

Fig. 5.8.    The Reichert TM 60 specimen trimmer. (By courtesy of C. Reichert AG.)

in turn for trimming. It is not possible to position the blade to cut across the top of the block to expose the tissue and so this has to be done on the ultramicrotome; nor is it possible to orientate the block in relation to the cutting tool. A lamp is fixed directly under the specimen block so that the light shines through it illuminating the specimen.

The Reichert TM60 specimen trimmer (Fig. 5.8) is a more sophisticated instrument consisting of a diamond milling cutter revolving at high speed. This is moved across the surface of the block and advanced towards it between each stroke, using a micrometer screw. The top of the block is first presented for trimming and then each of the sides in turn. An orientating head is used and adaptors are available which will accept specimen holders from most ultramicrotomes. The degree of slant of the sides of the block can be set to any angle and different shapes of block face can be produced, although there are pre-set positions for convenience. A stereomicroscope is fitted to the instrument together with a fixed focused light. The specimen holder is transferred to an ultramicrotome to cut survey sections and then returned to the specimen trimmer for the final trimming. A net micrometer eyepiece enables specific areas on the block face to be located by comparison with a stained survey section viewed under a microscope with a similar eyepiece.

The LKB Pyramitome is primarily a microtome incorporating a glass knife as the cutting edge (Fig. 5.9). The specimen arm is moved by a hand wheel and is automatically advanced towards the knife between each stroke. The advance is pre-set in increments of 1 µm, from 0 to 10 µm per stroke. A standard LKB specimen holder and orientation head are used so that they may be transferred directly to either of the LKB Ultrotomes. Adaptors for other ultramicrotome holders are also available. A stereomicroscope is mounted on the instrument for the observation and accurate alignment of the specimen and the knife and a focused light can be rotated about the cutting point. With the aid of the orientation head, and a centring disc fitted to the free end of the specimen arm, a selected area of the specimen block can be moved into the correct cutting position. The knife turret can accommodate three glass knives at a time, so that a fresh knife can be quickly rotated into position for cutting survey sections and further trimming. The knife stage is set perpendicular to the specimen arm and sections cut from the end of the block until the specimen is exposed. For trimming the sides the knife stage is rotated to any chosen angle to one side of its original position so that sections may be cut from the side of the block. The whole of the specimen head assembly is then rotated in a vertical plane to present each side in turn for trimming. The positions required to produce a trapezoidal or rectangular block face can be quickly located although any position is possible.

An accessory to this instrument, the Target Marker (Fig. 5.9), enables the image of a stained survey section to be superimposed on the block face (Kindel 1970). The Pyramitome is fitted with the Junger Micrall microscope and the normal objective lens is replaced by the Target Marker. A glass slide with a stained section is placed on the microscope stage of the Target Marker and when illuminated from below the image is reflected by a series of mirrors

Fig. 5.9.   The LKB Pyramitome shown with the Junger Micrall microscope and the Target Marker. (By courtesy of LKB Produkter AB.)

Fig. 5.10. Diagram showing how the image of a stained section is superimposed on the face of the specimen block with the aid of the Target Marker.
(By courtesy of LKB Produkter AB.)

to appear in front of the specimen block (Fig. 5.10). With the normal illumination the block face will also be seen, and the two images can be matched by adjusting the position of the glass slide. The desired area for re-trimming can then be accurately located on the block face.

## 5.4 The mesa technique

When scanning for specific structures throughout a specimen it is often desirable to take survey sections at regular intervals between the series of ultrathin sections. If the block is trimmed down in the normal way for the first series of ultrathin sections, much of the specimen will have been cut away and the subsequent survey section will have been considerably reduced in size, defeating the objective. Production of a mesa block (an isolated hill having steeply sloping sides and a level top) will avoid this problem. De Bruijn and McGee-Russell (1966) first described this technique and used a dental burr to form the mesa but a mesa can also be made with a glass knife using an ultramicrotome or the LKB Pyramitome. In principle any ultramicrotome can be used but problems may be encountered if the bypass movement of the specimen arm causes the specimen block to touch the back of the knife on the return stroke. After the first thick survey section has been taken (Fig. 5.11a) the corner of a straight-edged glass knife is placed to one side of the desired area. The knife is used in a similar manner to a chisel and sections a few micrometres thick are taken until a depth of 20 μm–30μm is reached (Fig. 5.11b). The specimen holder, or head, is then

Fig. 5.11.   Stages in the production of a mesa. (a) Survey section of the whole area. (b-e) Trimming the sides of the mesa. (f) Sectioning the mesa.

rotated through 90°, the corner of the knife moved to the desired position and the next side trimmed (Fig. 5.11c). The other two sides are then trimmed in a similar manner (Fig. 5.11d and e). To obtain a trapezoidal shape the specimen holder or head should be rotated 90° ± 5°, depending on which side is being trimmed, to produce the required angle. The resultant mesa can then be sectioned on the ultramicrotome (Fig. 5.11f) and when sufficient ultrathin sections have been obtained the mesa is cut away and another survey section of the whole area taken. This can be continued right through the specimen with the mesas being made at the same spot or in different positions throughout the specimen.

REFERENCES

De Bruijn, W. C. and S. M. McGee-Russell (1966), Bridging a gap in pathology and histology, Jl R. microsc. Soc. *85*, 77.
Fineran, B. A. (1971), A device for holding LKB-Ultrotome chucks during preliminary block trimming, J. Microscopy *94*, 83.
Galey, F. (1963), A mechanical technique for trimming tissue blocks in electron microscopy, J. Ultrastruct. Res. *9*, 139.
Glauert, A. M. (1974), Fixation, dehydration and embedding of biological specimens, in: Practical methods in electron microscopy, A. M. Glauert, ed. (North-Holland, Amsterdam).
Kindel, L. (1970), Rational preparation prior to ultrathin sectioning, Proc. 7th Int. Congr. Electron Microscopy, Grenoble *1*, 419.

# Chapter 6

# Sectioning procedures

## 6.1 The choice of a site for the ultramicrotome

Concentration is important for good sectioning so ideally the ultramicrotome should be situated in a quiet room or cubicle without distractions. Complete isolation is not desirable and a small room adjacent to the main laboratory will be a good position. Other work may then be carried out in between periods of sectioning to provide a relaxation. Most ultramicrotomes are affected to some degree by external vibrations and changes of temperature, so this should be taken into account when choosing a site. The instrument should not be placed in direct sunlight, near a draught such as an open window, or close to heavy vibrations. Some instruments are provided with their own tables; others may function well on an ordinary laboratory bench but they are less likely to be influenced by transmitted vibrations if they are placed on a separate sturdy table. Where there is a serious vibration problem, a special vibration damping table such as the Barry Company's Serva Bench Mark III, the Japanese Visolator or the LKB Ultrostable can be used (see Appendix). Good seating is essential for concentration. A chair with an adjustable seat height is desirable so that each operator can adjust it to a comfortable position.

## 6.2 Equipment required for sectioning

The following equipment should be brought to hand before starting sectioning to avoid interruptions. A small hair such as an eyelash is fixed to one end of a small stick with a drop of wax of glue. This is one of the most useful aids in ultramicrotomy as it can be used to manipulate the sections, remove unwanted ones, and to clean the knife edge without causing damage. To

288

clean the hair, it is normally sufficient to wipe it between a piece of folded lens tissue. A small bottle of the flotation liquid is preferable to a large supply which can easily become contaminated. A small bore syringe or pipette will be needed to transfer the liquid to the trough. There should be an adequate supply of specimen grids, a pair of fine tweezers for handling them and a grid box, or a Petri dish with two layers of filter paper in the bottom, in which to store them. Sufficient glass knives with troughs should be prepared for one session of sectioning, preferably with a selection of different angles if the sectioning properties of the specimen are not known. If a diamond knife is being used, a cleaning stick and some acetone or alcohol will be needed. Filter paper and lens tissues will also be found useful.

## 6.3 Adjustment of the knife

To avoid damaging the knife it is better to attach the specimen holder, or head, to the specimen arm before fixing the knife, although orientation of the specimen block will be made later. The knife stage should be withdrawn a reasonable distance before placing the knife in position so that there is adequate space between the knife edge and the specimen block to allow adjustments to be made. The knife height sets the cutting point and should ideally coincide with the mid-point of the specimen arm movement. In all present day ultramicrotomes the arm is moved through an arc and if the knife is set too high or too low it will alter the tangential relationship between the knife edge and the radius of the arm movement. Most instruments have a reference mark for setting the knife height but it is advisable to follow the instructions in the manufacturer's handbook. The knife should be clamped firmly but care should be taken to avoid breaking the seal between the knife and the trough. The clearance angle, between the knife and the specimen (Fig. 1.1) is normally set to 4° or 5° but will have to be varied depending on the properties of the specimen block. The knife stage is moved laterally to bring a suitable part of the knife edge in line with the face of the specimen block and is then rotated to bring the knife edge as near as possible to a perpendicular position in relation to the specimen arm; a scale is often provided for this purpose. If the knife edge is set obliquely to the specimen arm lateral distortions may appear in the sections similar to those described by Dempster (1942) in paraffin sections.

## 6.4 Adjustment of the specimen block

The specimen block should be firmly clamped in its holder to avoid any movement during sectioning. If the specimen block is soft it is advisable to allow the block to settle in its holder for 10–15 min before starting to section. Tightening the holder often causes such a specimen block to be slowly squeezed out of its holder and this would cause an undesirable movement towards the knife. This will often happen with freshly polymerized specimen blocks that have not hardened sufficiently. Before commencing any orientation, the fine and ultra-fine mechanical advance systems should be re-set to the beginning of their traverse. The specimen block is orientated so that the longer of the two parallel edges of the block face will hit the knife first. The specimen arm should be arrested at the mid-point of its movement whilst the height of the block face is adjusted to coincide with that of the knife edge. This is achieved either by raising or lowering the specimen head, or by adjusting the complete cutting stroke. The method will vary between different instruments and the manufacturer's instructions should be followed. To make it possible for the sections to adhere together forming a straight ribbon, the upper and lower edges of the block face must be parallel to the knife edge. The binocular microscope is positioned so that the front of the block face can be seen and the specimen head or holder rotated until this is achieved (Fig. 6.1). The block face must be parallel to both the knife edge and to the cutting plane if the whole of the surface is to be cut. If the specimen block has been machine trimmed or previously sectioned, the smooth face can be used to reflect an image of the knife edge. The light is positioned to obtain

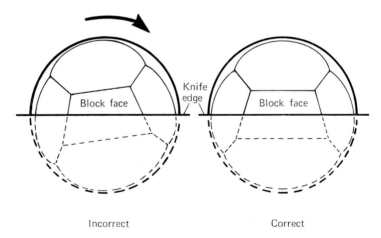

Incorrect                                   Correct

Fig. 6.1.   The specimen block is orientated so that the upper and lower edges of the block face are parallel to the knife edge. The broken lines indicate the part of the face that is behind the knife when viewed from the front.

a bright white reflection from the block face and the knife carefully advanced until its reflection is seen on the mirror-like surface of the block face; Gorycki (1965) describes a mirror attachment which he used for the same purpose. The specimen holder is moved, if adjustments are possible, in a horizontal plane until the knife edge and its reflection are parallel to each other (Fig. 6.2). It may be found to be easier to rotate the knife stage if only a small adjustment is needed but it should not be moved more than a few degrees from its perpendicular relationship to the specimen arm. When the specimen arm is moved slowly up and down the distance between the knife edge and its reflection should remain constant. Correction can be made by tilting the specimen holder in a vertical plane.

When the alignment is satisfactory the knife is carefully advanced until only a very small distance remains between the knife edge and its reflection; this indicates the gap between the knife and the block face. On some ultra-microtomes where the position of the lamp and the microscope are fixed, the positions of the light and dark reflections may be reversed. The gap between the knife and the block face will then appear dark instead of white as shown in Fig. 6.2. It is better to have too large a gap than risk damaging the cutting

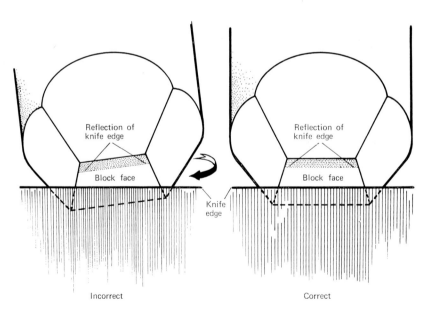

Fig. 6.2. To ensure that the whole of the block face is cut, the specimen holder is adjusted until the knife edge is parallel to its reflection seen in the mirror-like block face. (Viewed at a small angle from the vertical.)

edge by touching the block face. With practice it will be possible to reduce the gap to a few micrometres. Boeré (1971) describes a quick method for making this approach to within 0.5 μm. If difficulty is encountered in obtaining a bright reflection from the block face, the microscope should be set to a vertical position and the actual gap between the specimen block and knife observed (Fig. 6.3). This method however is not so precise as the one previously described as only the upper edge of the block face can be clearly seen. If the block face is very irregular or adjustments of the specimen holder are limited, sections will have to be taken until the full area of the block face is being cut; after facing the specimen block in this way a fresh knife edge should be aligned before commencing ultrathin sectioning. Although it may seem difficult at first, the more precise the alignment the quicker a good ribbon of sections will be obtained.

## 6.5   Trough liquids

For most purposes clean distilled water is perfectly adequate. Its physical properties make it almost ideal for this purpose; it has a relatively high surface tension so that the sections float and a low viscosity allowing them to move freely on its surface. The low evaporation rate enables the level to

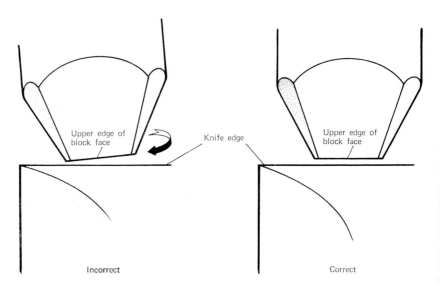

Fig. 6.3.   When viewed from above, the upper edge of the block face should be parallel to the knife edge.

be maintained without the need for frequent adjustments. In the early days of ultramicrotomy, when methacrylate embeddings were used extensively, it was usual to use a 10%–20% aqueous solution of ethanol or acetone as this helped to stretch out wrinkles in the sections. Although these solutions lower the surface tension they have little stretching effect on epoxy and polyester resin sections. Other solutions or liquids may be required when components in the specimen are likely to be affected by water, or when sectioning at low temperatures. Boothroyd (1964) used water saturated with calcium and phosphate ions to prevent demineralization of sections of embryonic bone, and Phillips (1961) suggests the use of xylol when sectioning metals that oxidise very rapidly. At low temperatures, ethylene glycol, ethylene alcohol, dimethyl sulphoxide, glycerol, freons and isopentane are some of the liquids that have been used. It is however difficult to find a suitable liquid which has the same advantages as water. When using an alternative, problems are often encountered because the sections sink instead of remaining on the surface; they are badly folded because they cannot move freely due to a high viscosity; or it is difficult to maintain the liquid at the knife edge without overfilling the trough. Little has been published on the subject of trough liquids and if water cannot be used it will probably be a case of trial and error until a suitable liquid is found for a particular specimen.

Before filling the trough it is advisable to raise or lower the specimen arm to prevent the block face being wetted should the trough be accidentally overfilled. Using a syringe or a pipette the liquid is added at the rear of the trough, to avoid touching the knife edge, until the knife edge is completely wetted and the liquid surface is approximately horizontal. When difficulty is encountered in wetting the knife edge, as is often found with a new diamond knife, more liquid than necessary should be added and then slowly removed until the desired level is reached. Failing this the liquid can often be drawn up to the knife edge with the aid of a small hair moistened with saliva or a suitable wetting agent. The fluorescent light is positioned to obtain a silvery white reflection over as large an area of the liquid surface as possible but particularly in the region where the sections will be formed. If the microscope is tilted a few degrees from the vertical towards the operator the normal position of the lamp will also be a few degrees from the vertical but in the opposite direction. Small adjustments are now made to the liquid level to obtain the optimum reflection from the surface. A misshapen knife or badly mounted trough will cause shadows and reduce this area considerably. Against this white background variations in the thickness of the sections, within a certain range, are seen as different colours (§6.7). Sections should always be of an even thickness and therefore it is important to be able to see when changes in the colour occur and vary the cutting conditions immediately (Fig. 6.4). If ultrathin sections appear colourless then the correct reflection has not been obtained. It may not be possible to obtain such a reflection when liquids other than water are used.

The knife edge must always be kept wet. If the level of the liquid is too low the edge will quickly become dry as evaporation occurs. If the meniscus is too high and therefore convex, the liquid may be dragged over the back of the knife when sectioning commences. With instruments that have an adjustable lamp, small variations in the liquid level can be compensated for by moving the lamp. Where the light source is fixed the liquid level must

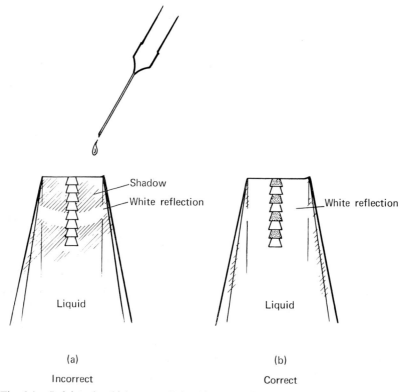

(a)                                   (b)

Incorrect                           Correct

Fig. 6.4. Judging the thickness variations in the sections by viewing them in reflected light. (a) Only a small area of the liquid surface has the correct reflection so the varying thickness of the sections is not easily seen. This appearance will normally indicate that the liquid level is too low, but a similar view will be seen if the meniscus is highly convex. (b) The correct reflection reveals that the sections vary in thickness.

be adjusted. This is normally done by using a fine bore syringe or pipette, although a finer control can be obtained if a level control device such at that described by Westfall and Healy (1962) or the Reichert Reflexomat (see Appendix), is used.

If the liquid is not perfectly clean it should be discarded as the dirt is likely to be picked up with the sections. Contamination may come from the trough itself, the filling device, or the liquid storage bottle. If necessary the liquid should be filtered before use.

## 6.6   The production of sections

The first contact between the specimen block and the knife is most important as both can be damaged if care is not taken. The specimen arm is operated by either the manual or automatic control and the knife advanced in very small increments, approximately 1 μm, between each stroke. It is impossible to judge accurately by eye the distance between the block face and the knife edge and if the advance is too great there is a danger of cutting a very thick section which will blunt the cutting edge. The choice of cutting speed will vary for different specimens (see §6.8) but the initial speed should not be too slow or the liquid will be dragged over the back of the knife before the first section is cut. An average speed of 2–3 mm per second can be used as a starting point and the speed then varied according to the properties of the material being sectioned. When cutting soft specimen blocks a faster speed than is desirable may be required to detach the first section and once sectioning has commenced the speed can be reduced.

Once contact has been made the ultra-fine feed is set and the specimen arm movements continued automatically, or by hand (with hand operated instruments) in a steady rhythmic manner. While sectioning is in progress attention should be paid to avoiding any disturbances that are likely to affect it. Any adjustments that have to be made on the ultramicrotome should be done gently and care taken not to bump the table or bench. As the advance takes place automatically between each stroke it is important that a section is produced at every stroke once contact has been made, otherwise the specimen block will rub against the clearance facet of the knife thus quickly damaging the knife edge. When starting with a new specimen it is advisable to begin with an advance of approximately 90 nm, or gold sections, and then decrease this as quickly as possible in steps of about 5 nm until a suitable thickness is reached (§6.7). The sections should be allowed to float freely on the liquid surface. If the knife edge is dry or the leading edge of the ribbon is impeded, the sections will be creased (Fig. 7.1a). Sectioning is continued until sufficient sections have been obtained or until sectioning becomes erratic. The arm movement and the advance are then stopped and the sections collected. It may be possible to continue sectioning on the same part of a glass knife edge but it will no longer be as sharp, so it is far better to move immediately to a fresh part of the knife edge or to a new knife. Some realignment may be necessary if the knife edge is not perfectly straight but this can be done just as easily with the trough full, by one of the methods already described. With a diamond knife sectioning can be continued on the same part of the edge but it is advisable to retract the knife stage 1–2 $\mu$m when sectioning is stopped. The action of cutting compresses the front of the specimen block slightly and this will expand during the rest period. If the knife is not retracted a thick section will be cut when sectioning is recommenced possibly causing damage to the diamond. The formation of the sections should be observed carefully as this will give some indication of the consistency of the specimen block. Wrinkles which appear as the section is cut but slowly disappear indicate a softness, whilst sections from a firm block will always be perfectly smooth.

## 6.7   Choice of section thickness

The smallest section thickness that can be obtained will depend on the size of the block face, the quality of the embedding and the sharpness of the cutting edge. The ultra-fine advance takes place automatically between each stroke and if a section is not produced at one stroke the following section will be approximately twice as thick as the advance setting. If the advance is set below the level that the knife is capable of cutting, the block face will be compressed by rubbing against the clearance facet of the knife. On the next stroke the advance, together with the expansion of the block, will be sufficient to cause a section to be detached, although thicker than desired. The advance should be increased until sections are produced regularly at every stroke, or a sharper knife must be used. If the edge of a glass knife has been allowed to dry after once being wet it will no longer be capable of producing good sections and so should be discarded.

When sections floating on the surface of the water are viewed through the binocular microscope interference colours are seen. Light is reflected from both the upper surfaces of the sections and from the section–water interface and these light waves, moving out of phase, produce interference colours related to the thickness of the sections and the refractive index of the

# TABLE 6.1.

The relation between thickness and colour of sections for an embedding medium of refractive index 1.5. (See Peachey 1958.)

Continuous interference colour and
thickness scale for thin sections

Thickness

Colour

n: index of
refraction = 1.5

— Grey

nm

60 —

— Silver

90 —

— Gold

150 —

— Purple

190 —

— Blue

240 —

sectioned material. These colours have been described by Peachey (1958) who produced a colour scale (Table 6.1); this scale is applicable to all embedding materials having a refractive index close to 1.5, including methacrylates, epoxy resins and polyester resins. The colours form a continuous spectrum and are not distinct steps so these values can only be taken as a guide to the actual thickness (see Chapter 9). The thinner sections are more difficult to cut and handle and it is inadvisable to aim for the thinnest possible unless this is absolutely essential to the experiment (Chapter 1). Most sections within the silver to pale gold range (60 nm–90 nm) are suitable for normal work, although thinner ones will be required when high resolution is needed.

## 6.8 Methodical sectioning

When floating on the trough liquid the sections should appear uniformly coloured both within each section and from section to section in the ribbon. The surface of the sections should be smooth without wrinkles or folds, which are seen as dark lines approximately parallel to the knife edge. When viewed in the electron microscope the sections should appear of an even thickness i.e. without chatter (see §7.9) and structures should not be deformed due to compression (see §7.7). Apart from the characteristics of the specimen block and the sharpness of the cutting edge there are three main variables which affect the quality of the sections, the cutting speed, the clearance angle and the actual knife angle. When a new or difficult specimen is to be cut a systematic test should be made to find the optimum conditions; a plan for methodical sectioning is shown in Table 6.2. A knife with a medium angle (60°) is selected and set to a medium clearance angle (4°). Three short ribbons of sections are then cut, each at a different cutting speed (2, 1 and 5 mm per second). Sections that appear perfect should be collected for more critical examination in the electron microscope. Higher and lower values of clearance angle are then selected and the experiment repeated. The same process can then be repeated with knife angles of higher and lower values. The actual knife angles should vary by at least 5° if any noticeably different result is to be seen, and the clearance angle by 2° or 3°. Unfortunately as specimens vary so considerably no rules can be given for changing these values. The following guide lines can be followed but if no improvement is found it is worthwhile making changes in the opposite direction. The main difficulty is the fact that there is no standard specimen block with which comparisons can be made.

Fast cutting speeds normally increase the amount of compression and

## TABLE 6.2
### A plan for methodical sectioning

| Test number | I | II | III |
|---|---|---|---|
| Knife angle | M(edium) 60 | H(igh) 65 | L(ow) 55 |
| Clearance angle | M 4°  L 1°  H 7° | M 4°  L 1°  H 7° | M 4°  L 1°  H 7° |
| Sectioning speed | M L H<br>2 1 5<br>mm/s mm/s mm/s<br><br>M L H<br>2 1 5<br>mm/s mm/s mm/s<br><br>M L H<br>2 1 5<br>mm/s mm/s mm/s | M L H<br>2 1 5<br>mm/s mm/s mm/s<br><br>M L H<br>2 1 5<br>mm/s mm/s mm/s<br><br>M L H<br>2 1 5<br>mm/s mm/s mm/s | M L H<br>2 1 5<br>mm/s mm/s mm/s<br><br>M L H<br>2 1 5<br>mm/s mm/s mm/s<br><br>M L H<br>2 1 5<br>mm/s mm/s mm/s |

The values shown cover the sectioning of most types of biological specimen. Each test is made with a knife of a different angle (actual, not scored, angle) and three different cutting speeds tried at each of three different clearance angles. (By courtesy of LKB Produkter AB.)

folding. Too slow a speed results in no section being cut but liquid being dragged over the back of the knife. Hard resin blocks generally cut better at slow speeds (1 mm per second or less) but very hard specimens, such as metals and bone, may require high speeds (10 mm per sec or more). If chatter is found when sectioning a hard block the sectioning angle should normally be decreased, when present in a softer block it should be increased. This may mean a change in either the knife angle or the clearance angle, or both. It is often believed that the knife angle should be as small as possible but in fact actual knife angles of 60° to 65° will often produce much better results than smaller angles if the specimen block is not too hard. Such a knife usually has a long usable edge and the edge appears to retain its sharpness longer. Once the optimal conditions have been found the values can be used with similarly embedded specimens until problems arise, when further tests should be made.

## REFERENCES

Boeré, H. (1971), Methods for orienting and aligning the pyramid front-face and knife edge on the LKB Ultrotome ultramicrotomes I and III, Science Tools (LKB) *18*, 24.

Boothroyd, B. (1964), The problem of demineralisation in thin sections of fully calcified bone, J. Cell Biol. *20*, 165.

Dempster, W. T. (1942), Distortions due to the sliding microtome, Anat. Rec. *84*, 269.

Gorycki, M. A. (1965), Aligning block faces by reflected light for precise sectioning, Stain technol. *40*, 265.

Peachey, L. D. (1958), Thin sections. I. A study of section thickness and physical distortion produced during microtomy, J. biophys. biochem. Cytol. *4*, 233.

Phillips, R. (1961), Diamond knife ultramicrotomy of metals and structure of microtomed sections, J. appl. Phys. *12*, 554.

Westfall, J. A. and D. L. Healy (1962), A water control device for mounting serial ultrathin sections, Stain technol. *37*, 118.

Chapter 7

# Sectioning problems

With well prepared specimens and good knives ultramicrotomy is not difficult; the skill comes in being able to recognise a problem and knowing the remedy. The following are the most common problems that are likely to be encountered but be prepared to find more than one fault present at a time.

## 7.1 Irregular sectioning

If no sections are produced after the first contact has been made a check should be made to see that the ultra-fine advance is operational and has not reached the end of its travel. If sections are produced but only at each alternate stroke, or vary in thickness from section to section, the cutting edge is not sharp enough to cut the thickness selected. This may be due to a blunt knife edge, which can easily be changed, or to the quality of the embedding. In this case the specimen arm advance should be increased until sections of an even thickness are produced at each stroke. If they are then too thick to be of value the area of the block face should be reduced. This problem will often be seen when the specimen block is very soft as the tip of the block is displaced during sectioning and so is out of position when the next cut is made. There will be less tendency for this to happen if the sides of the block are not trimmed too steeply. Soft blocks can sometimes be hardened by leaving the trimmed block in an oven at 60 °C overnight.

Irregular thickness can also be caused if the knife and the specimen block are not firmly clamped in their holders, or if there are random movements of the specimen arm. These may be in the direction of the advance or lateral displacement between strokes (Glauert and Phillips 1965).

## 7.2 Poor ribbon formation

If the sections do not adhere together in a ribbon the most usual reason is that the upper and lower edges of the block face are not straight. This can be caused by trimming with a damaged blade so that ridges are formed on the sides of the block resulting in an irregular edge at the block face. When the upper edge of the block face is not parallel to the knife edge each section will only contact the following section at one point and so easily becomes detached (Fig. 7.1b). If the liquid level is too high the meniscus at the knife edge may cause the sections to leave the edge immediately they have been cut so preventing any contact between them.

The ribbon will curve towards one side if the top and bottom edges of the block face are not parallel to each other (Fig. 7.1a) or if one side of the sections is more compressed than the other due to uneven sharpness of the

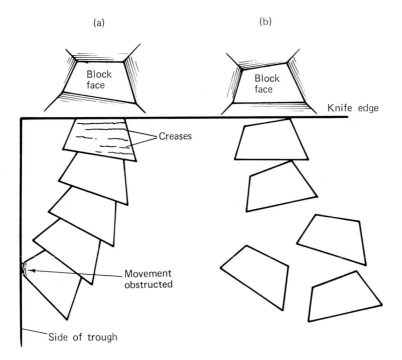

Fig. 7.1. The upper and lower edges of the block face are not parallel to one another. (a) The ribbon of sections has curved towards the side of the trough and become obstructed. (b) When the top edge of the block is not parallel to the knife, the sections are only in contact with each other at one point and quickly float apart.

cutting edge. This is more often seen using a diamond knife where parts of the edge are used more than others. It is important to ensure that the leading end of the ribbon is not obstructed by touching the side of the trough or other impediment otherwise the sections will not be able to float out smoothly but will become creased (Fig. 7.1a).

### 7.3   Sections dragging over the knife edge

This often occurs with a soft specimen block because the section is not cleanly detached from the upper edge of the block face, but is torn away leaving a ragged edge. Faster cutting speeds will help to detach the section more cleanly and a slight lowering of the liquid level will prevent the section overhanging the knife edge due to a high meniscus. It is, however, important to ensure that the whole block face completely passes the knife edge before the specimen arm starts upon its return stroke. An increase in the clearance angle and/or knife angle may also help to overcome this problem.

### 7.4   Wetting of the block face

If the liquid level is too high liquid will be drawn over onto the back of the knife by the specimen as it passes. Once this happens the block face must be blotted dry and any liquid remaining on the clearance facet of the knife carefully removed with a piece of tissue or filter paper before recommencing sectioning. When sectioning large areas, more than 2 $mm^2$, it is often necessary to reduce the liquid level as far as possible without allowing it to recede from the knife edge. With smaller areas the level is not normally so critical. Some specimen blocks do not cut well at slow speeds; they rub against the clearance facet of the knife drawing the liquid over the knife edge. When this happens the knife should be cleaned by wiping a fine hair across the back of the knife edge to detach any embedding material that has adhered to it and then a faster cutting speed should be selected. Once sectioning has started it may be possible to reduce the speed if necessary to prevent compression.

### 7.5   Holes in the sections

This is really a problem of poor embedding and little can be done to improve conditions at this stage. The problem generally arises where there are areas of hard tissue in the specimen e.g. calcified structures, or where natural boundaries in the specimen prevent good impregnation of the embedding medium, e.g. cuticle layers in botanical specimens. It is not possible to soften a hard specimen during the embedding process but by varying the proportions of the embedding components, a harder block can be produced which should correspond to the hardest part of the material (Glauert 1974). This will give a more even consistency to the specimen block and make sectioning much easier.

## 7.6 Scratches

Any straight mark running through a series of sections perpendicular to the knife edge will have been produced by a defect in the edge (Fig. 7.2) (Glauert and Phillips 1965). Most defects can clearly be seen by critical examination of the edge, as described in §3.5, and so should be avoided. If scratches are observed on the sections a fresh part of the edge or a new knife should imme-

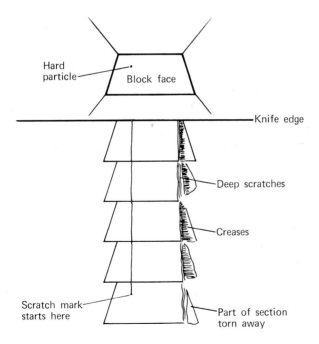

Fig. 7.2. The right-hand side of the ribbon of sections shows deep scratches resulting in tearing and creasing of the sections. As the scratch on the left does not appear in the leading part of the first section the damage has been caused by a hard particle within the specimen.

diately be tried. The first section or two may still contain the score marks as the fault in the knife may have damaged the block face; if however the scratch continues to persist in exactly the same place on the sections the fault will lie with the specimen. A hard region within the specimen can damage the knife edge at its first contact which thereafter produces a scratch mark. The damage may be caused by a foreign body which can be trimmed away, but if it is inherent in the tissue e.g. asbestos particles in lung tissue, the problem may only be solved by resorting to a diamond knife. A similar result may be seen where small particles of the material are torn out from areas that have not been well impregnated with the embedding medium. These particles often adhere to the knife edge causing scratch marks on subsequent sections until they are dislodged (Fig. 7.3) (Kindel and Sevéus 1971).

Few sections are completely free from scratches and in fact one or two can be helpful in indicating the direction of cutting and therefore the direction of any compression. Occasionally the sections appear covered by fine scratches when viewed at about 5000 × in the electron microscope (Fig. 7.4). They are caused by a poor quality knife edge. The apparently good part of the knife edge does have very small irregularities although they cannot be seen except at very high magnifications. Normally they do not cause any problem but if they do give rise to scratches the amount of apparently usable edge should be increased by better control in the way the knife is broken, and only that part furthest away from visible defects used. These scratches may easily be confused with chatter (§7.9) and therefore it is important to note the relationship of the marks to the leading edge of the section; scratches will be perpendicular to the knife edge and chatter parallel to it.

## 7.7   Compression

As stated in Chapter 1, the ideal section would have the same area as the block face from which it was detached. There is normally no difference between the width of the section and the width of the block face, but the length of the section is usually shorter; it is this shortening that is referred to as compression. Wrinkles and folding of the section are often associated with compression. These are seen as dark wavy lines on the section approximately parallel to the knife edge but do not necessarily extend the full width of the section. When examined in the electron microscope badly compressed sections will exhibit an altered structural orientation in the direction of cutting (Fig. 7.5).

The use of vapours of organic solvents or heat to stretch out this compres-

Fig. 7.3. The first part of each of a series of sections is free from scratches. An area containing calcium has not been well impregnated with the embedding medium and has therefore been torn out. Small particles have adhered to the knife edge causing scratch marks only in the latter part of the sections. As each fresh section is cut the particles are dislodged from the edge so that the scratches only re-appear after the calcium area has been cut. (a) Region of scratches; (b) material torn out; (c) direction of cut. (From Kindel and Sevéus 1971.)

sion is sometimes advocated (Sotelo 1957; Satir and Peachey 1958). A small artist's brush or a strip of filter paper is soaked in a suitable solvent and then held over the sections. The brush or paper must be held very close to the surface of the sections for the heavy vapours to have any effect, but care must be taken to avoid actually touching the sectioning or they will be destroyed. Xylol can be used for stretching methacrylate sections but epoxy resins sections are much more difficult to expand. Chloroform and ethylene chloride will, however, normally produce some expansion. Heat can be applied in a

Fig. 7.4.   Fine scratches covering the whole of the section can easily be confused with chatter unless their relationship to the direction of cut is studied. Scratches are parallel to the direction of cut (arrow), and therefore perpendicular to the leading edge of the section.

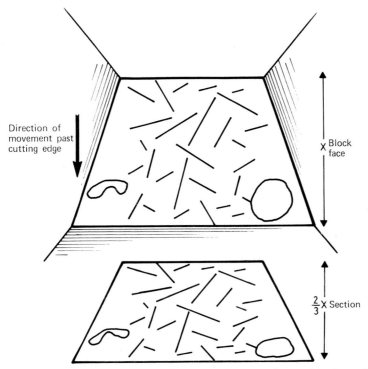

Direction of
movement past
cutting edge

X Block
face

$\frac{2}{3}$X Section

Fig. 7.5.   The section shows a considerable amount of compression resulting in an altered
orientation of the structures in the direction of cutting.

similar way by bringing a hot wire filament into close proximity to the sections. Roberts (1970) describes a heat pen for this purpose (see Appendix).
The sections will be seen to expand suddenly and the heat or vapour should
then be removed to prevent overstretching which would cause structures to
split apart. Expansion is accompanied by a decrease in section thickness
which can be seen from the change in interference colour. Cosslett (1967)
has shown that this stretching of sections can give rise to even greater variations in thickness within the section than normally occurs. Although no
section is completely free from compression it is far better to limit compression by the correct choice of cutting conditions, than to try to remove it
later. Sections from soft specimen blocks have a greater tendency to compress
than those from hard blocks.

### 7.8   Variations in thickness within a section

Varying thickness within a section can have three possible causes. If the demarcation between the two thicknesses is perpendicular to the knife edge it indicates that the cutting edge is not equally sharp in both regions. It is possible that one side of the section will only be cut at every second stroke. Selecting a better cutting edge should cure this fault. If the demarcation runs irregularly through the section the fault will be caused by the uneven consistency of the specimen block (Fig. 7.6). This is often seen between the

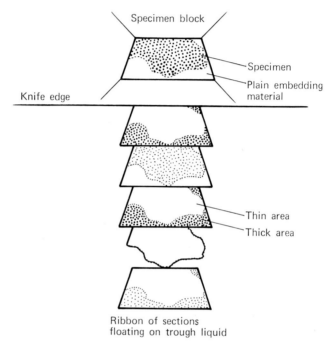

Fig. 7.6. Variation in thickness within a section caused by the uneven consistency of the specimen block. The demarcation line runs irregularly through the sections following the edge of the specimen or some structure within it.

specimen and plain embedding material. Re-trimming is the only solution but attempts should be made to improve the quality of the embedding in the future. If the thickness variations are in the form of bands running parallel to the knife edge (Fig. 7.7) they are usually caused by vibrations, between the knife and the specimen block, from the instrument, or from external sources. Careful observation through the microscope of a shadow on the surface of the trough liquid when the instrument is not in operation will show whether the vibrations are from an external source. If the vibration is

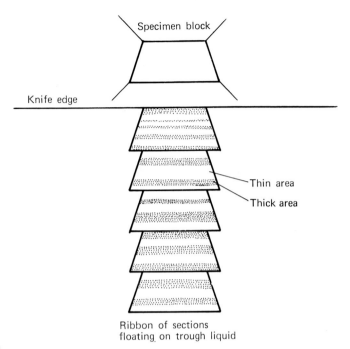

Fig. 7.7. Variations in thickness seen as straight bands parallel to the knife edge are caused by vibrations.

continuous re-siting of the instrument or its isolation on a special table will be necessary. It should be remembered that disturbances can often be caused by the operator knocking the table or the microscope during operation of the instrument. If the cause does not appear to be external, first check that the knife and specimen block are firmly clamped. Vibrations may be set up if a hard specimen block hits the knife edge at too fast a speed; also there is a greater tendency for vibrations to be set up if the leading edge of the block face is very wide. Similar vibrations are caused when using a diamond knife at an incorrect clearance angle; the angle may need to be increased or decreased. If the problem still persists it would indicate a fault in the instrument.

## 7.9  Chatter

The term 'chatter' is used to describe the effects caused by vibrations set up in machining tools but in electron microscopy it has come to mean the effect

Fig. 7.8. 'Chatter' seen as closely-spaced variations in section thickness perpendicular to the direction of cut (arrow) which can be identified by the scratch mark.

of a high frequency vibration set up between the knife and the specimen (Phillips 1962). The frequency is too high to involve any component of the ultramicrotome. The result is a closely spaced variation in section thickness parallel to the knife edge, with a wave length of 1 μm or less, and is therefore only seen in the electron microscope (Fig. 7.8). This vibration is only set up under certain conditions and changes in the sectioning angle and cutting speed will often overcome it. Chatter will sometimes only appear in certain regions of the specimen; this is due to variations in the composition of the material being cut causing some parts to deform more than others during the cutting process. The solution may have to be a compromise; conditions being selected that give reasonable but not the best results from all areas, or the microscopist must be satisfied to obtain good results from one region at the sacrifice of others.

REFERENCES

Cosslett, A. (1967), Letters to the editor, Jl R. microsc. Soc. *86*, 315.
Glauert, A. M. (1974), Fixation, dehydration and embedding of biological specimens, in: Practical methods in electron microscopy, A. M. Glauert, ed. (North-Holland, Amsterdam).
Glauert, A. M. and R. Phillips (1965), The preparation of thin sections, in: Techniques for electron microscopy, 2nd ed., D. H. Kay, ed. (Blackwell Scientific Publications, Oxford), p. 213.
Kindel, L. and L. Sevéus (1971), Scratches in ultrathin sections appearing in electron micrographs, LKB Application Note, May 1971.
Phillips, R. (1962), Comment on localized, short spaced periodic variation in contrast on methacrylate sections, Jl R. microsc. Soc. *81*, 41.
Roberts, I. M. (1970), Reduction of compression artifacts in ultra-thin sections by the application of heat, J. Microscopy *92*, 57.
Satir, P. G. and L. D. Peachey (1958), Thin sections. II. A simple method for reducing compression artifacts, J. biophys. biochem. Cytol. *4*, 345.
Sotelo, J. R. (1957), Technical improvements in specimen preparation for electron microscopy, Expl Cell Res. *13*, 599.

# Chapter 8

# Collection of the sections

Having produced the sections they must now be placed on a specimen grid for viewing in the electron microscope. It is very disheartening to have taken so much care to produce a good ribbon of sections only to destroy or damage them during collection. It calls for steady hands and some practice. Unwanted sections should first be removed so that they do not cause an obstruction when manipulating those that are required. They can be removed quite easily by firmly prodding them one at a time, with a single hair to which they will stick. The hair must then be carefully wiped between folded lens tissue to remove the debris before touching the good sections. The ribbon of sections can be detached from the knife edge by touching the top of the last section with the clean hair and pulling it gently away from the knife edge. The ribbon can then be moved easily into the centre of the trough where there is more room for manipulation and less risk of damaging the cutting edge. If the ribbon is too long for one specimen grid it can be divided by a sharp tap with the hair on the border between two sections. If this does not succeed two separate hairs should be used to pull the sections apart (Fig. 8.1). The sections should only be touched on their upper surface; if attempt is made to move the sections by touching their outer edges, they are likely to adhere to the wet hair.

## 8.1 Specimen support grids

The specimen grid should give adequate support to the sections without obscuring too great an area. Copper grids are normally chosen, except where chemical treatment of the sections may affect them; then, nickel, gold, stainless steel or platinum grids are used. Many different types of grid are now

Fig. 8.1. A ribbon of sections being divided by two fine hairs.

available (see Appendix), and a small selection is shown in Fig. 8.2. The mesh size refers to the number of openings to the linear inch and for most routine work the 200 or 300 mesh size will be adequate. Smaller mesh values may be needed when the sections are very small and would not receive adequate support from the grid bars. The hexagonal type gives more support to very thin sections and the single hole grids are used to give an uninterrupted view of a ribbon of sections. Identification marks on the bars enable areas to be quickly located in the electron microscope. The surfaces of the grids generally have a different appearance, one side matt and the other shiny. The matt surface is slightly rougher and is often selected for the mounting of the support film and sections. However, both sides appear to be equally good, although it is advisable always to use the same side as it is difficult to see to which side the sections are attached once they have dried.

If the sections are large enough to cover several of the grid bars no additional support is usually needed but with small sections, or with single hole grids, a supporting film is required. This is normally made by floating a thin film of collodion or formvar (approx. 20 nm thick) on a water surface. Clean grids are then placed on the floating film and covered with a piece of clean paper. Once this has been wetted the paper carrying the grids and the film is removed carefully and allowed to dry. The grids are then ready for use. A thin carbon film is sometimes used either alone or on top of a plastic film to provide a conducting layer which helps dissipate the heat from the electron beam. Further details of the preparation of support films can be found in Bradley (1965) and Goodhew (1972). Although a film gives support to the sections it also reduces the contrast and if not well made it may

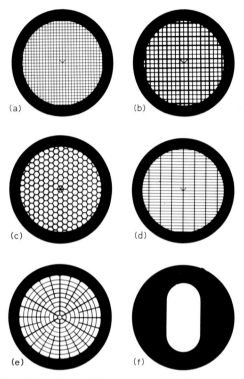

Fig. 8.2.   A selection of specimen grids: (a) 300 mesh; (b) 200 special; (c) hexagonal; (d) 75 × 300; (e) radial; (f) single hole, 1 mm × 2 mm. (By courtesy of Polaron Equipment Ltd.)

disintegrate in the electron beam, causing the sections to be lost and also contaminating the electron microscope. Grids will normally be clean enough to use directly from their container but if necessary they can be washed in alcohol. Used grids can be cleaned (Bradley 1965) but it is usually not worth the effort unless they are an expensive type; invariably some debris remains which interferes with future viewing.

## 8.2   Methods of collecting the sections

A suitable specimen grid is picked up in a pair of fine tweezers (No. 7) equipped with a small rubber ring to hold the points together. Curved tweezers may be found to be easier to use but this is a matter of individual preference. If the sections have formed a straight ribbon the grid is lowered vertically into the liquid a little way behind the sections until it is just submerged. The ribbon is then gently pulled towards the grid with the aid of a perfectly clean fine hair. When the edge of the first section lies over the rim of the grid, the grid is lifted vertically out of the water (Fig. 8.3). This method of collection sacrifices part of the first section on the rim of the grid, but prevents folding of the rest of the sections. If the upper edge of the grid is tilted away from the sections and the grid is withdrawn slightly from the liquid it is possible to position the ribbon below the rim of the grid but the ribbon will be more difficult to attach to the grid. In this position the grid causes a break in the reflection from the liquid surface making it more difficult to see the

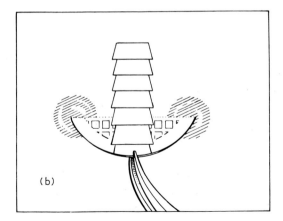

Fig. 8.3. (a) A specimen support grid held in a pair of tweezers. (b) The grid partly removed from the water with one end of the ribbon of sections adhering to it.

sections and also the grid tends to repel the ribbon of sections, particularly when a supporting film is being used. If the sections have not formed a ribbon, but are floating independently on the liquid surface, the above method cannot be used. A grid is then placed horizontally underneath the sections and then slowly raised, lifting out the sections on a drop of water. As the tweezers must enter the trough from above it is necessary to bend the rim of grid upwards where it is held in the tweezers, in order to place it in the correct position (Fig. 8.4). This must be done carefully so as not to damage the support film. The grid should be raised slowly through the liquid or the sections will float away from it. The grid may also be lowered down onto the sections from above so that they are picked up together with a small drop of the liquid. It is very difficult to place the grid absolutely flat onto the sections with the result that they are often creased when using this technique.

After collecting the sections, the grid can be blotted dry by touching its edge to a piece of filter paper or left to dry while still held in the tweezers. If the grid is released while it is still wet, the surface tension of the liquid will cause

Fig. 8.4.   A cluster of single sections being collected by raising the grid underneath them.

the grid to remain adhering to one of the points of the tweezers with possible damage to the sections. Once the sections have dried they remain firmly fixed to the grid and can be transferred to staining solutions (Lewis et al. 1974). When collecting sections from trough liquids other than water it may be necessary to rinse the grid after collection of the sections in a suitable liquid, e.g. alcohol, to remove any residual liquid which may cause contamination in the electron microscope.

When examining serial sections it is important to have an uninterrupted view of the whole ribbon. Single hole or slot grids are generally used for this purpose. As there are no grid bars to support the sections some type of supporting film must be used. This may need to be a little bit thicker than normal to prevent it splitting under the electron beam with a consequent loss of the sections. The ribbon of sections must be carefully positioned over the hole or slot and Gay and Anderson (1954) developed a special technique for achieving this. A loop, somewhat larger than the specimen grid, is made at the end of a short length of thin wire. This is covered by a formvar film in a similar way to that described for covering grids. The ribbon of sections is then lifted out of the trough using the loop in place of a grid. A slot grid is then placed on the upper end of a perspex rod fixed to the condenser mount of a microscope. The loop is orientated above the grid until the sections lie over the slot. The grid is then raised through the loop so that the formvar film and the sections are transferred to the grid. Barnes and Chambers (1961) describe a simple micromanipulator made from glass rods and microscope slides for orientating the loop with the sections over the grid.

Sections may also be removed from the trough using an uncoated wire loop (Noble 1962), or a plastic annulus (Marinozzi 1964) or an uncoated single hole grid (Galey and Nilsson 1966). The loop, annulus or grid is lowered onto the trough liquid directly over the sections and immediately removed. The sections will now be found floating on the surface of the droplet of liquid remaining on the loop, and may be transferred directly to a staining solution (McGee-Russell 1966) or to another site for more precise orientation for mounting on the specimen grid. Anderson and Brenner (1971) transferred the sections to liquid gelatin which was then cooled immobilising the sections. A grid is then carefully positioned over them and the gelatin re-melted by placing in an oven at 60 °C. When the grid is removed the sections remain attached to it. Unwanted gelatin is removed by floating the grid on a 2% acetic acid bath for 30 min, followed by 3–5 min on Tris buffer, pH 7.1, to neutralize, and then 30 min on distilled water to wash. All procedures are carried out at 60 °C.

Fig. 8.5.   A specimen grid being lowered into the trough with the aid of the LKB Section
Collector. (By courtesy of LKB Produkter AB.)

## 8.3 Mechanical aids to section collection

Additional handling of the sections increases the risk of damage to them. More accurate positioning of the sections on the grid can be made directly when collecting them from the trough if a micromanipulator or a water control device is used. Westfall (1961) and Westfall and Healy (1962) cemented a bakelite wedge to the sectioning facet of a knife to provide a flat horizontal surface about 1 mm below the knife edge on which a specimen grid is placed. After a ribbon of sections has been cut they can easily be manipulated into position over the grid with the aid of a fine hair and then the trough liquid is slowly withdrawn by the control device lowering the sections onto the grid. Behnke and Rostgaard (1964) described a 'third hand' with which the grid can be positioned while leaving both the operator's hands free for manipulating the sections. The LKB Section Collector is a development of this device in which a pair of tweezers holding a specimen grid is clamped on an arm attached to the microscope of the Ultrotome (Fig. 8.5). The grid can be fixed at any angle in relation to the liquid surface by rotating the clamp so that any of the collection methods described in §8.2 can be used. The top of the grid is brought into the focal plane of the microscope by moving the arm in a vertical plane with the aid of a screw thread. Once the grid is aligned it remains in focus all the time and can be gently lowered into the trough by lowering the complete microscope with the aid of the microscope focusing knob. Once contact has been made with the sections the grid is raised clear of the trough in a similar manner whilst observing the operation through the microscope.

Sections mounted on grids should be stored in a dust-free container whilst awaiting examination in the electron microscope. As several grids will be prepared at one sectioning session a good identification system is necessary. There are several grid storage boxes available that fulfil these needs and are excellent for long term storage (see Appendix).

REFERENCES

Anderson, R. G. W. and R. M. Brenner (1971), Accurate placement of ultrathin sections on grids; control by sol-gel phases of a gelatin flotation fluid, Stain technol. *46*, 1.
Barnes, B. G. and T. C. Chambers (1961), A simple and rapid method for mounting serial sections for electron microscopy, J. biophys. biochem. Cytol. *9*, 724.
Behnke, O. and J. Rostgaard (1964), Your 'third hand' in mounting serial sections on grids for electron microscopy, Stain technol. *39*, 205.
Bradley, D. E. (1965), The preparation of specimen support films, in: Techniques for electron microscopy, 2nd ed., D. H. Kay, ed. (Blackwell Scientific Publications, Oxford), p. 58.

Galey, F. R. and S. E. G. Nilsson (1966), A new method for transferring sections from the liquid surface of the trough through staining solutions to the supporting film of a grid, J. Ultrastruct. Res. *14*, 405.

Gay, H. and T. F. Anderson (1954), Serial sections for electron microscopy, Science N.Y. *120*, 1071.

Goodhew, P. J. (1972), Specimen preparation in materials science, in: Practical methods in electron microscopy, A. M. Glauert, ed. (North-Holland, Amsterdam).

Lewis, P. R., D. P. Knight and M. A Williams (1974), Staining methods for thin sections, in: Practical methods in electron microscopy, A. M. Glauert ed. (North-Holland Amsterdam).

Marinozzi, V. (1964), Cytochimie ultrastructurale du nucléole-RNA et protéines intra-nucléolaires, J. Ultrastruct. Res. *10*, 433.

McGee-Russell, S. M. (1966), A ring and cup technique for handling thin sections for electron microscopy, Jl R. microsc. Soc. *85*, 91.

Noble, H. W. (1962), A rapid and accurate method of mounting sections for electron micro-scopy, Jl R. microsc. Soc. *80*, 295.

Westfall, J. A. (1961), Obtaining flat serial sections for electron microscopy, Stain technol. *36*, 36.

Westfall, J. A. and D. L. Healy (1962), A water control device for mounting serial ultrathin sections, Stain technol. *37*, 118.

# Chapter 9

# Measurement of section thickness

For a full interpretation of the specimen under observation to be made it is necessary to know the thickness of the sections. This is of particular importance when three-dimensional reconstruction, stereology or autoradiography is undertaken (see Lewis et al. 1974). The advance setting of an ultramicrotome cannot be relied upon to indicate the section thickness as it does not take into account variations due to random thermal changes and compression during sectioning. Unfortunately at the present time there is no quick reliable method of measuring section thickness. The interference colours seen when the sections are floating on water provide a rough guide to the section thickness, but interpretation of these colours varies with individuals. Even with the most careful estimations it has been shown that there can be great variations, not only between sections of the same colour but within the same section (Williams and Meek 1966; Cosslett 1967). There is an instantaneous decrease in section thickness on exposure to the electron beam even when the beam current is kept low. Ideally therefore thickness measurements should be made before and after irradiation to allow for this. Sjöstrand (1953) and Porter and Blum (1953) calculated the section thickness by measuring the shadow lengths after shadowing with a heavy metal at a known angle. The disadvantage of this method is that the measurement only refers to the extreme edge of the section and gives no indication as to whether the thickness is even throughout. Peachey (1958) prepared his colour scale (Table 6.1) from thickness measurements made by ellipsometry. Other workers (Zelander and Ekholm 1960; Cosslett 1960) made measurements using the interference microscope but found it difficult to measure accurately in the range most useful to the electron microscopist, 30 nm to 80 nm. More recently Gillis and Wibo (1971) have used interferometry in conjunction with a photo-multiplier and claim an accuracy of 0.1 nm with ultrathin sections. Williams and Meek (1966) used specimen blocks containing a radioisotope and by measuring the area of the sections and their total radioactivity the thickness of the sections could be calculated. A considerable variation in thickness was seen within each section (see Lewis et al. 1974). All these techniques are concerned with measuring the section thickness outside the electron microscope. Other methods involve the analysis of the electron micrographs.

The thickness may be calculated by observation of an internal standard of known dimensions within the section when it is viewed in the electron microscope. Huxley (1957) was able to make use of a naturally occuring standard, the regular spacing of muscle filaments, but this is rarely possible.

Silverman et al. (1969) embedded stained spheres of Dowex anion exchange resin to provide a thickness standard. Ultrathin sections were cut and the optical transmission of the sectioned spheres measured. The mass and density were calculated to give the volume and, by measuring the area, the thickness of the sections could be calculated. Mota (1960) and Phillips and Shortt (1964) embedded sections mounted on a grid and then re-sectioned across the thickness of the grid. Measurements were made in the electron microscope of the original section thickness. Weibull (1970) describes a method for calculating the thickness based on an estimation of the relative number of electrons that are scattered when the electron beam is passing through the section. If the microscope is equipped with a high-angle tilting stage assessment of the thickness of the section can be made in certain areas (Willis 1971). The stage is tilted until a structure traversing the thickness of the section is seen in maximum contrast, i.e. it is orientated parallel to the electron beam, and the angle of tilt is recorded. By tilting the specimen through 90° the extended width of the object can be measured. The specimen thickness is given by this width multiplied by the cosine of the first tilt angle.

None of these techniques are quick enough to be considered for routine use, and so normally statements regarding the thickness of sections should only be considered as approximations unless otherwise stated.

REFERENCES

Cosslett, A. (1960), The effect of the electron beam on thin sections, Proc. 2nd. Eur. Reg. Conf. Electron microscopy, Delft 2, 678.
Cosslett, A. (1967), Letters to the editors, Jl R. microsc. Soc. *86*, 315.
Gillis, J. M. and M. Wibo (1971), Accurate measurement of the thickness of ultrathin sections by interference microscopy, J. Cell Biol. *49*, 947.
Huxley, H. E. (1957), The double array of filaments in cross-striated muscle, J. biophys. biochem. Cytol. *3*, 631.

Lewis, P. R., D. P. Knight and M. A. Williams (1974), Staining methods for thin sections, in: Practical methods in electron microscopy, A. M. Glauert ed. (North-Holland, Amsterdam).

Mota, M. (1960), Sectioning of sections for determination of thickness in electron microscopy, Melhoramento *13*, 127.

Peachey, L. D. (1958), Thin sections. I. A study of section thickness and physical distortion produced during microtomy, J. biophys. biochem. Cytol. *4*, 233.

Phillips, R. and T. Shortt (1964), The 're-sectioned section' technique and its application to studies of the topography and thickness of thin sections, Jl R. microsc. Soc. *82*, 263.

Porter, K. R. and J. Blum (1953), A study in microtomy for electron microscopy, Anat. Rec. *117*, 685.

Silverman, L., B. Schreiner and D. Glick (1969), Measurement of thickness within sections by quantitative electron microscopy, J. Cell Biol. *40*, 768.

Sjöstrand, F. S. (1953), A new microtome for ultrathin sectioning for high resolution electron microscopy, Experientia *9*, 114.

Weibull, C. (1970), Estimation of the thickness of thin sections prepared for electron microscopy, Philips Bulletin EM 45.

Williams, M. A. and G. A. Meek (1966), Studies on thickness variation in ultrathin sections for electron microscopy, Jl R. microsc. Soc. *85*, 337.

Willis, R. A. (1971), *In situ* measurement of EM section thickness by use of a high angle tilting stage, J. Anat. *109*, 345.

Zelander, T. and R. Ekholm (1960), Determination of thickness of electron microscopy sections, J. Ultrastruct. Res. *4*, 413.

Chapter 10

# Examination of ultrathin sections in the electron microscope

Before starting to examine the sections in the electron microscope check that there are sufficient photographic plates or films ready for use in the camera and evacuation chamber.

Ultrathin sections are easily damaged by the electron beam in the electron microscope and consequently considerable care should be taken when viewing the sections and taking the final records on photographic plates or film. The careful selection and application of the appropriate fixation, embedding (Glauert 1974), sectioning and section staining (see Lewis et al. 1974) techniques is to no avail if the sections are then damaged at the final stage.

All embedding media are modified by electron bombardment *in vacuo*; material is lost, leading to decreases in thickness and total mass of the section. The actual amount of material lost depends on the thickness of the section and the conditions of exposure to the electron beam, but the loss of mass is always less for the epoxy and polyester resins (13% to 30%) than for the methacrylates (40% to 60%) (see review by Reimer 1965). In addition, the loss of mass from methacrylate sections is accompanied by a flow of the remaining material leading to the disruption of fine structure (Morgan et al. 1956). This loss can be reduced by covering the sections with a carbon layer so that they are then 'sandwiched' between two films (Watson 1957). The carbon layer can be evaporated onto the section before it is placed in the electron microscope, or the sections can be scanned at low beam intensity without an anti-contaminator in operation so that a carbon layer is deposited on the sections as contamination. The methacrylate sections are then more stable but the additional carbon layers results in a loss of contrast and resolution.

Epoxy and polyester resin sections are more stable under electron bombardment and their three-dimensional structure appears to be preserved.

Plastic support films are also modified by electron bombardment, collodion losing far more mass (65% to 85%) than formvar (15% to 40%) (Reimer 1965). Consequently an unstable specimen may result from damage to the support film, even if the section is relatively stable. Plastic films can be made more stable by coating them with a thin layer of evaporated carbon. This layer protects both the support film and the section. Instability of the plastic-carbon film may still cause problems and sections should be mounted on pure carbon films, or mounted on naked grids, for work at high resolution.

The changes in sections and support films occur instantaneously as soon as the beam is switched on. Consequently it is advisable to scan a grid square at low beam intensity (and low magnification) at first to allow the section and film to stabilize as much as possible. Check that the condenser has been left fully overfocused before switching on the beam. The beam intensity should then be continued to be kept as low as possible throughout the examination of the specimen until sufficient photographic recordings have been obtained. It is often advisable to photograph an area by the 'minimum contrast' method (Agar et al. 1974) at low illumination before increasing the illumination to view the section through the binocular telescope.

The anti-contaminator should always be used, except when depositing a protective layer as described above, to prevent a layer of carbon building up on the section and reducing contrast. Even so it is advisable to take photographs as soon as possible in order to obtain the maximum contrast.

It is not advisable to scan a section in one electron microscope and then examine it in greater detail in a higher resolution instrument. Specimen damage is usually greatest in small and medium resolution instruments, and also a contaminating layer will be deposited on the sections during transfer at atmospheric pressure. For the same reason sufficient photographs should be taken at the first examination of the sections as the contrast will be progressively reduced each time the section is re-examined.

Photographic plates and films should be numbered and details of the specimen and magnification recorded immediately. It is surprisingly easy to forget these details the next day even if only one or two photographs are taken.

*Reid   Ultramicrotomy*

REFERENCES

Agar, A. W., R. H. Alderson and D. Chescoe (1974), Principles and practice of electron microscope operation, in: Practical methods in electron microscopy, A. M. Glauert, ed. (North-Holland, Amsterdam).

Glauert, A. M. (1974), Fixation, dehydration and embedding of biological specimens, in: Practical methods in electron microscopy, A. M. Glauert, ed. (North-Holland, Amsterdam).

Lewis, P. R., D. P. Knight and M. A. Williams (1974), Staining methods for thin sections, in: Practical methods in electron microscopy, A. M. Glauert, ed. (North-Holland, Amsterdam).

Morgan, C., D. H. Moore and H. M. Rose (1956), Some effects of the microtome knife and electron beam on methacrylate-embedded thin sections, J. biophys. biochem. Cytol. *2*, suppl. p. 21.

Reimer, L. (1965), Irradiation changes in organic and inorganic objects, Lab. Invest. *14*, 1082.

Watson, M. L. (1957), Reduction of heating artifacts in thin sections examined in the electron microscope, J. biophys. biochem. Cytol. *3*, 1017.

# Chapter 11

# Special techniques

## 11.1   Three-dimensional analysis

Sections only present a two-dimensional view of structures but by re-combining them in their original order a three-dimensional image can be obtained. This involves cutting a large number of serial sections and photographing the particular structure under observation in each section. The photographic plates can then be stacked in sequence and a composite photograph taken through them (Birch-Andersen 1955; Fuscaldo and Jones 1959). Alternatively a model of the structure under study is made from each section in a material of proportional thickness and the models are then stacked in sequence on top of each other (Bang and Bang 1957; Sjöstrand 1958). Pedler and Tilly (1966, 1968) used a pantograph; structures are traced in each electron micrograph and simultaneously the corresponding shapes are cut out of sheets of polystyrene by a hot wire (Fig. 11.1). These shapes are then stacked together to form the model. Mitchell and Thaemert (1965) obtained three-dimensional information by drawing the structures seen in electron micrographs in perspective by a modification of graphic reconstruction. This can be compared very simply to drawing the sides of a railway track converging at a point to create the appearance of depth. Tracings are made of the relevant structures from micrographs of serial sections. A three-point perspective chart can be purchased which facilitates the construction of the perspective drawing of the structures in each section. After allowing for the thickness of the sections these drawings can be combined into one picture showing the complete structure. Yamada and Yoshida (1972) constructed two stereo maps by tracing the contours of the structures from each section onto maps with a small lateral displacement between each section. The maps

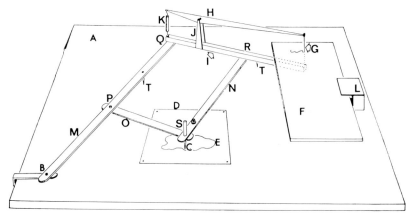

Fig. 11.1   Diagram showing the general arrangement of the hot wire pantograph. (A) Baseboard; (B) teflon bearing; (C) stylus; (D) micrograph; (E) profile to be traced; (F) expanded polystyrene sheet; (G) electrically heated wire; (H) tapered bar; (I) rigid bar; (J) fulcrum joint; (K) tension spring; (L) clamp holding polystyrene; (M, N, O) strip metal arms; (P, Q, R, S) Teflon bushed joints; Teflon pads. (From Pedler and Tilly 1966.)

were then studied stereoscopically. Dunn (1972) describes a similar method. By tracing part of the contours of the structures onto a common plan with a small lateral displacement between each section, a clear three-dimensional image is obtained. This method, although not as accurate as others described, is relatively simple.

It is essential with these techniques that the sections are kept in sequence and the minimum number of sections lost, if a realistic view is to be obtained. As it may be necessary to examine several hundred sections a diamond knife is a great advantage to avoid the numerous changes that would have to be made if glass knives were used. A series of sections is cut and the ribbon then divided into lengths suitable for each specimen grid. These can be parked at the side of the trough before being collected on grids in the correct order (Fig. 8.1).

McCulloch (1965) describes a method for obtaining two separate ribbons of sections, each being the right-angled complement of the other, which when examined side by side on the same grid make it possible to observe the third dimension along the line of the common edge.

Tilting stages in the electron microscope make it possible to examine stereo images of sections giving a three dimensional view, and with high voltage electron microscopes, where sections 0.5 μm and more in thickness can be examined in this way, a much better interpretation of the third dimension can be obtained (see report of Symposium on Applications of High Voltage Electron Microscopy, EMCON 72, J. Microscopy *97*, parts 1 and 2, 1973.)

## 11.2 Sectioning methods for non-biological material

Although many non-biological materials have been successfully sectioned by ultramicrotomy (Goodhew 1972) they are generally more difficult to handle than biological specimens. Most do not embed well because it is difficult or impossible to impregnate them with the embedding monomer. This results in the embedding material only adhering to the outer surfaces of the specimen and it is nearly always torn away during sectioning or during examination in the electron microscope. If the specimen has sufficient mechanical strength and a large enough sample is available it is advisable to clamp it directly into the specimen holder of the ultramicrotome without embedding. Trimming of very hard specimens such as metals can be done by milling or grinding; the block face should be kept as small as possible, approximately $0.1 \text{ mm}^2$. During trimming the amount of material removed from the specimen at each cut should become less as the desired area is reached or the tip of the specimen may become excessively worked. Very porous material, powders and fibres must be embedded in a resin whose hardness should match, as near as possible, that of the specimen.

Considerable deformation may occur during sectioning, such as dislocations (R. Phillips 1961; V. A. Phillips 1960, 1962) and it is thought that the procedure is more a controlled fracture than actual slicing with brittle materials (V. A. Phillips 1967). However, as long as this is appreciated useful information can still be obtained. Some of the softer materials can be cut with a glass knife but generally a diamond knife will be found to be necessary. Difficulty is often encountered in obtaining ribbons of sections or even whole sections with such specimens but this is not too great a problem as the structure is generally more uniform than in biological specimens and it is always worthwhile examining the sectioned pieces as useful information may be obtained from them. Sections of unembedded material will rarely appear as smooth as those of embedded specimens and generally will not exhibit any interference colours; any colours that are seen are difficult to interpret due to the different refractive indices. Certain materials such as fibres and crystal structures will often section more easily in one direction than others and so the specimen should be rotated in relation to the knife edge until the optimum cutting plane is found. A fibre presented parallel to the cutting edge will be

much more difficult, if not impossible, to cut. If it is rotated through 90° so
that it is perpendicular to the cutting edge a much more even strain will be put
on the knife (see §5.1). With hard or difficult materials sectioning will often
be found to be easier if the block face is presented obliquely to the knife or
made triangular in shape so that one of the apices touches the knife edge
first. In this way the important first impact with the cutting edge is kept as
small as possible although it will be impossible to obtain a ribbon of sections

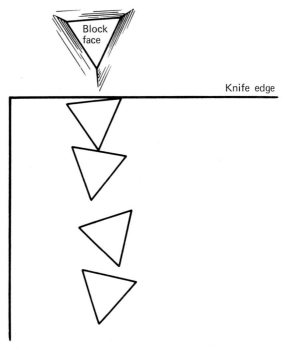

Fig. 11.2.   Triangular-shaped block face presented obliquely to the knife edge to offer
minimum resistance at the first contact.

(Fig. 11.2). Slip lines are often formed when cutting metals and can be seen
on the upper surface of the section (Phillips and Shortt 1964); they are related
to the sectioning angle. If the sectioning angle (Fig. 1.1) is too large there is a
greater shear strain on the sections which will result in wrinkles and folds.

With certain metals, problems of oxidation may be encountered when
using water as a trough liquid and xylol has been suggested as a suitable
alternative. It will then be necessary to use another trough sealant that will

not be dissolved by xylol. Other liquids may be required when the specimen is affected by water (see §6.5). Some of the materials that have been successfully sectioned are aluminium, cadmium, magnesium, zinc, copper, nickel, gold, beryllium, steel, various alloys, glass, carbon, clay and many polymers. Goodhew (1972) compares the technique of ultramicrotomy with other methods of thinning specimens in materials science.

REFERENCES

Bang, B. G. and F. B. Bang (1957), Graphic reconstruction of the third dimension from serial electron microphotographs, J. Ultrastruct. Res. *1*, 138.
Birch-Andersen, A. (1955), Reconstruction of the nuclear sites of *Salmonella typhimurium* from electron micrographs of serial sections, J. gen. Microbiol. *13*, 327.
Dunn, R. F. (1972), Graphic three-dimensional representations from serial sections, J. Microscopy *96*, 301.
Fuscaldo, K. E. and H. H. Jones (1959), A method for the reconstruction of three-dimensional models from electron micrographs of serial sections, J. Ultrastruct. Res. *3*, 1.
Goodhew, P. F. (1972), Specimen preparation in materials science, in: Practical methods in electron microscopy, A. M. Glauert, ed. (North-Holland, Amsterdam).
McCulloch, J. R. (1965), 'Three-dimensional' ultramicrotomy, J. Ultrastruct. Res. *13*, 34.
Mitchell, H. C. and J. C. Thaemert (1965), Three dimensions in fine structure, Science, N.Y. *148*, 1480.
Pedler, C. and R. Tilly (1966), A new method of serial reconstruction from electron micrographs, Jl R. microsc. Soc. *86*, 189.
Pedler, C. and R. Tilly (1968), A source of error in the interpretation of electron micrographs, Jl R. microsc. Soc. *88*, 177.
Phillips, R. (1961), Diamond knife ultramicrotomy of metals and structure of microtomed sections, J. appl. Phys. *12*, 554.
Phillips, R. and T. Shortt (1964), The 're-sectioned section' technique and its applications to studies of the topography and thickness of thin sections, Jl R. microsc. Soc. *82*, 263.
Phillips, V. A. (1960), Structure of ultramicrotomed multi-phase alloys, Proc. 2nd Eur. Reg. Conf. Electron Microscopy, Delft *1*, 485.
Phillips, V. A. (1962), The deformation of metals and alloys by ultramicrotomy, in: Direct observation of imperfections in crystals, J. B. Newkirk and J. H. Wernick, eds (Wiley and Sons, New York and London), p. 179.
Phillips, V. A. (1967), Use of a diamond knife ultramicrotome in light and electron microscopy on metals, Prakt. Metallog. *4*, 637.
Sjöstrand, F. S. (1958), Ultrastructure of retinal rod synapses of the guinea pig eye as revealed by three-dimensional reconstructions from serial sections J. Ultrastruct. Res. *2*, 122.
Yamada, M. and S. Yoshida (1972), Graphic stereo-reconstruction of serial sections, J. Microscopy *95*, 249.

Chapter 12

# Cryo-ultramicrotomy – ultrathin sectioning at low temperatures

## 12.1 Biological specimens

Cryo-ultramicrotomy, or ultracryotomy as it has been called, cannot really be considered as a routine technique at the present time. However, the development shown over the last few years has been very encouraging and indicates that the technique will be much more widely used in the future. Much of what has already been said on the techniques for ultramicrotomy also applies to cryo-ultramicrotomy, but in addition there are other factors to be taken into account. First let us consider what are the advantages of cryo-ultramicrotomy when the technique is perfected. The first is rapid results; the normal embedding procedure takes several days before the tissue is ready to be sectioned and examined in the electron microscope. In cryo-ultramicrotomy results can be obtained much more quickly, as the lengthy dehydration and embedding procedures are avoided. In the ideal method the fresh tissue is frozen rapidly without further treatment and is then sectioned at low temperature. This means that the chemical composition of the tissue should remain unaltered and soluble compounds will not be extracted. Frozen sections then provide an ideal medium for histochemical and immunocytochemical reactions, soluble compound autoradiography and X-ray micro-analysis.

Although cryo-ultramicrotomy was first attempted on biological specimens as early as 1952 by Fernández-Morán little success was reported until the work of Bernhard (1965) and Stumpf and Roth (1965). These two groups of workers used the technique of placing a complete ultramicrotome inside a refrigerated cabinet or cryostat. Later workers (Appleton 1968; Yamataka 1972) have also used this method. Advantages of the cryostat method are good temperature stability and a large working area. Although cooling to the required temperature takes several hours, once this temperature has been reached it is relatively stable due to the heavy mass of the ultramicrotome. Cooling is achieved by a compressor system and this can be left running for days or even weeks before defrosting is necessary. Disadvantages are the high cost of such an instrument and the fact that temperatures lower than − 70 °C are difficult to achieve with standard methods of refrigeration. The ultramicrotome will normally have to be modified for work within the cryostat, and this makes it difficult to use as a conventional ultramicrotome.

Most of the later work has been done with cryokits, that is with an attachment fitted to the knife and specimen area of a conventional ultramicrotome to cool only those areas. Cryokits are available for the LKB Ultrotomes, for the Reichert ultramicrotomes and for the Sorvall Porter-Blum MT2 ultramicrotome. Detailed descriptions of these instruments can be found in Persson (1972a, b) on LKB, Dollhopf et al. (1972) on Reichert, and Christensen and Paavola (1972) on Sorvall. With a cryokit cooling is normally achieved by means of liquid or gaseous nitrogen, and temperatures down to approximately − 170 °C can be obtained. This system is simpler and therefore less expensive than the cryostat, and by removing the cryokit, a relatively simple operation, the ultramicrotome can be used in the conventional manner at room temperature.

Both glass and diamond knives can be used at low temperature and it has been reported by several workers that the quality of a glass knife appears to improve under these conditions. Although the diamond itself appears to be unaffected by extremes of temperature, the fixing of the diamond within the metal shaft may become loosened after repeated cooling and re-warming of the knife. Sectioning is often carried out on a dry knife, that is without any floatation liquid present. If wet sectioning is preferred a tape trough can be attached to the glass knife as described in §3.7. It is, however, preferable to use a thin tape, for example Scotch polyester film tape No. 850, in order to have good thermal conductivity between the glass knife and the metal jaws of the knife holder. Wax can be used as a sealant with temperatures down to − 50 °C with some trough liquids, but leakage will generally occur if the cooled knife and trough liquid are subsequently brought to room temperature. Silicone glue can be used as a sealant at most temperatures and with most trough liquids, although it takes a little longer to set than wax.

Preparation of the specimen has been the subject of much discussion.

Some workers advocate the use of light fixation with glutaraldehyde followed by encapsulation in gelatin (Bernhard and Nancy 1964), methyl cellulose (Iglesias et al. 1971) or cross-linked albumin (Kuhlmann and Viron 1972). These encapsulating media (Glauert 1974) consist of macromolecules which do not penetrate the cells, but are used to give support to the tissue while it is being cut. Encapsulation can be followed by a short soaking in various concentrations of glycerol or dimethyl sulphoxide (DMSO) which gives some protection against the formation of ice crystals within the tissue during freezing. Tokuyasu (1973) describes a method of infusing the tissue with a chemically inert hydrophilic substance of low molecular weight, such as sucrose, which it is claimed improves the sectioning qualities and also gives protection against freezing artefacts. Sevéus (1969) omits both encapsulation and protective agents, while Appleton (1968) and Christensen (1969) have advocated the complete avoidance of all solutions. In this way they hope to preserve the chemical composition of the tissue in the most lifelike state.

The freezing of the specimen is done as quickly as possible. This can be achieved by plunging a small piece of tissue into a liquid with a low freezing point. The most commonly used substance is liquid nitrogen which is inert, but its freezing and boiling points are relatively close together and consequently gas bubbles may form around the tissue slowing down the cooling rate. Liquid helium has similar properties, but is rather more expensive. To overcome the formation of a gas layer some workers have frozen the tissue in isopentane which is itself cooled by liquid nitrogen; however, a layer of isopentane remains around the tissue and this sometimes causes difficulties in sectioning. Propane and Freon 22 have also been used as coolants, and Appleton (1974) recommends a sludge bath of solid nitrogen and liquid nitrogen to give faster cooling rates. The tissue is either placed directly on the specimen holder or on a small silver or copper pin, and is then plunged directly into the coolant so that the tissue freezes onto the holder or pin. The small mass of the pin allows a faster cooling rate and the pin can then be transferred directly into the specimen holder whilst still in the coolant. The tissue may be placed in a small droplet of water or methyl cellulose on the holder to ensure better fixing. Christensen (1969) prefers to freeze the tissue by bringing it into contact with a copper surface which has been cooled by liquid nitrogen.

Trimming can be carried out under the coolant using a sharp blade or *in situ* on the ultramicrotome using a glass knife or a trimming device (Hodson and Marshall 1969). The actual temperature at which the sections should be cut is another matter for debate, and it seems likely from the work reported so far that different tissues will require different temperatures. During sectioning some of the energy involved is converted into heat which raises the temperature of the sections and the block face (Persson 1972b). Hodson and Marshall (1972) calculated that a temperature rise of between 40 °C and 100 °C may occur, although so far no direct measurements on the tissue itself have been possible. From practical experience most workers have found that specimen temperatures between − 80 °C and − 100 °C give the best results. Some workers have reported that chipping rather than sectioning occurs at temperatures below − 100 °C, although this is not borne out by the experience of all workers. The temperature of the knife is limited by the freezing point of the trough liquid, if wet sectioning is employed. Several workers have used an aqueous solution of DMSO; a 60% solution will freeze at approximately − 60 °C. DMSO is, however, undesirable for cytochemical work as it inhibits acid phosphatases (Pfeifer et al. 1974) and may have other effects. Other liquids that have been used are ethylene glycol, methylene, glycerol, Freon, isopentane and liquid nitrogen. The trough liquid should have a surface tension and viscosity which are suitable for floating the sections at the desired temperature. Unfortunately none of the liquids listed above are ideal in this respect at temperatures below − 50 °C with the result that sections either sink below the surface or tend to pile up on top of one another because they cannot float freely. Sectioning on a dry knife eliminates these problems and allows a wider range of temperatures to be selected, but has the disadvantages that the sections are not flat and are more difficult to collect. Hellström (1973) describes an anti-roll plate for use in dry sectioning to give flatter sections, and Christensen (1969) flattens the sections once they are on the specimen grid by pressing with the smooth polished end of a copper rod.

The thickness of frozen sections is much more difficult to judge as they lack the interference colours seen in resin sections. Some colours are occasionally seen in thicker sections but as these have not been related to actual thickness measurements, they do not have much use at the present time. The only way the thickness can be estimated is by first ensuring that a stable temperature has been reached and that a section is produced at every stroke of the specimen arm. The feed calibration of the ultramicrotome can then be used as an approximate indication of the thickness of the sections.

Collection of the sections from a liquid surface can be achieved by touching a coated grid to the surface of the sections, or by collecting them within a small plastic loop as described by Marinozzi (1964) and then transferring them to the surface of a grid (see §8.2). Excess trough liquid, particularly in the case of DMSO, must be removed by washing immediately or it will interfere with the examination of the sections.

Collection of sections from a dry knife can be done by lifting the sections up with a small hair and depositing them on a coated specimen grid, or by direct contact, using a coated grid of a temperature slightly higher than that of the sections. This latter method, however, is liable to damage the knife edge. Appleton (1972) uses a small vacuum probe to support the leading section in mid-air while further sections are formed in ribbon formation. The ribbon is then gently lowered onto the surface of a grid. Tokuyasu (1973) uses a small drop of a saturated sucrose solution picked up on the tip of an eyelash probe. When introduced into the cryokit the droplet remains in a molten state long enough to be brought into light contact with the frozen sections. The droplet is then removed to room temperature and once melted the sections are transferred to the surface of a specimen grid or wash bath.

Some workers have recommended freeze-drying the sections before removal from the cryokit; however, results so far do not show any advantages for this method. Sections can be stained either positively or negatively or not at all.

With the development of equipment for X-ray micro-analysis a great deal of interest has been shown in ultrathin frozen sections, particularly of unfixed and untreated material (Hodson and Marshall 1970; Appleton 1972; Chandler 1972; Weavers 1973). The use of conventionally-embedded sections for X-ray micro-analysis is rather limited due to the presence of 'foreign' elements within the embedding matrix and the removal of elements from the tissue by the dehydrating and embedding processes (Davies and Erasmus 1973). A comprehensive list of published work in the field of cryo-ultramicrotomy has been prepared by LKB and is available from them on request.

## 12.2   Polymers and rubbers

Many polymers and rubbers are too soft to be cut unsupported at room temperature. Conventional embedding in a resin often causes swelling or changes in the structure of the specimen. An alternative method is to lower the temperature of the specimen below its glass transition point. One problem is that most of these materials are mixtures of different compounds, and therefore their exact glass transition point is not easily determined. Chappuis and Robblee (1957) first described the use of a freezing attachment for an ultramicrotome to cut carbon loaded rubbers using solid carbon dioxide and alcohol as the coolant. Other workers used liquid nitrogen as the coolant and obtained temperatures ranging from $-50°C$ to $-170°C$ (Leigh-Dugmore 1961; Hess and Ford 1963; Andrews and Stubbs 1964). In 1966 Crudgington described the forerunner of the LKB cryokit. The specimen, a rubber-like material, was cooled by contact with a reservoir of solid carbon dioxide and alcohol. A gentle stream of nitrogen gas was directed onto the knife and specimen to add further cooling and to prevent the formation of ice crystals. Further work has been reported on the sectioning of polyethylene by Andrews et al. (1967) and Dlugosz and Keller (1968). Cobbold and Mendelson (1971) have cut several types of polymers, both of bulk material and granules embedded in ice. Their results gave some indication that the sections produced were the result of controlled fractures rather than true cutting.

It has generally been found that materials will section better if the temperature is lowered well below the glass transition temperature of the material. This is probably necessary in order to counteract the heat produced by the cutting process. As DMSO reacts with many of the polymers this cannot be used as a floatation medium. Various mixtures of ethyl alcohol and ethylene glycol have been used, but the choice of a suitable floatation medium will depend on the actual material being sectioned and will have to be found by trial and error.

The rate of cooling of the specimen and the rate of warming of the sections after cutting do not appear to be critical, and the main problems facing the worker studying polymers and rubbers are the selection of the correct temperature and cutting conditions to give optimum results. The sections are collected in the same way as sections of frozen biological material.

## REFERENCES

Andrews, E. H., M. H. Bennett and A. Markham (1967), Macrostructure of bulk polymer as revealed in ultrathin sections, J. Polymer Sci. *5*, 1235.

Andrews, E. H. and J. M. Stubbs (1964), A new freezing head for the ultramicrotomy of rubbers, Jl R. microsc. Soc. *82*, 221.

Appleton, T. C. (1968), Ultrathin frozen sections for electron microscopy, Preliminary report, LKB Instruments Inc., Rockville, Maryland, U.S.A.

Appleton, T. C. (1972), 'Dry' frozen sections of unfixed and unembedded biological material for X-ray microanalysis of naturally occurring diffusible electrolytes: the cryostat approach, J. Microscopie *13*, 144.

Appleton, T. C. (1974), Cryo-ultramicrotomy, possible applications in cytochemistry, Proc. 2nd Int. Symp. Electron Microscopy and Cytochemistry, Drienerloo, 1973, (North-Holland, Amsterdam) p. 299.

Bernhard, W. (1965), Ultramicrotomie à basse température, Annls Biol. *4*, 5.

Bernhard, W. and M.-T. Nancy (1964), Coupes a congélation ultrafines de tissu inclus dans la gélatine, J. Microscopie *3*, 579.

Chandler, J. A. (1972), Recent developments in analytical electron microscopy, J. Microscopy *98*, 359.

Chappuis, M. M. and L. S. Robblee (1957), Improved sectioning technique for electron microscopy of carbon black stocks, Rubber World *136*, 391.

Christensen, A. K. (1969), A way to prepare frozen thin sections of fresh tissue for electron microscopy, in: Autoradiography of diffusible substances, L. J. Roth and W. E. Stumpf, eds. (Academic Press, New York and London), p. 349.

Christensen, A. K. and L. G. Paavola (1972), The Sorvall-Christensen frozen thin sectioner, J. Microscopie *13*, 147.

Cobbold, A. J. and A. E. Mendelson (1971), Ultrathin sectioning of polymers at low temperatures for electron-microscopical investigation, Science Tools (LKB) *18*, 1.

Crudgington, J. R. (1966), A freezing head for the ultra-thin sectioning of rubber-like materials, Science Tools (LKB) *13*, 23.

Davies, T. W. and D. A. Erasmus (1973), Cryo-ultramicrotomy and X-ray microanalysis in the transmission electron microscope, Science Tools (LKB) *20*, 9.

Dlugosz, J. and A. Keller (1968), Beam-induced contrast in the electron-microscope images of polymers as observed in ultra-thin sections, J. appl. Phys. *39*, 5776.

Dollhopf, F. L., G. Lechner, K. Neumann and H. Sitte (1972), The cryo-ultramicrotome Reichert, J. Microscopie *13*, 152.

Fernández-Morán, H. (1952), Application of the ultrathin freezing-sectioning technique to the study of cell structures with the electron microscope, Arkiv Fysik *4*, 471.

Glauert, A. M. (1974), Fixation, dehydration and embedding of biological specimens, in: Practical methods in electron microscopy, A. M. Glauert, ed. (North-Holland, Amsterdam).

Hellström, S. (1973), An antiroll plate for use when cutting frozen ultrathin sections on dry knives in the cryoultramicrotome, J. Ultrastruct. Res. *44*, 431.

Hess, W. M. and F. P. Ford (1963), Microscopy of pigment-elastomer systems (Columbian Carbon Co., New Jersey).

Hodson, S. and J. Marshall (1969), A device for trimming frozen biological tissue preparatory to ultra-thin sectioning, J. Microscopy *89*, 373.

Hodson, S. and J. Marshall (1970), Tissue sodium and potassium: direct detection in the electron microscope, Experientia *26*, 1283.

Hodson, S. and J. Marshall (1972), Evidence against through-section thawing whilst cutting on the ultracryotome, J. Microscopy *95*, 459.

Iglesias, J. R., R. Bernier and R. Simard (1971), Ultracryotomy; a routine procedure, J. Ultrastruct, Res. *36*, 271.

Kuhlmann, W. and A. Viron (1972), Cross-linked albumin as supporting matrix in ultra-thin cryo microtomy, J. Ultrastruct. Res. *41*, 385.

Leigh-Dugmore, C. H. (1961), Microscopy of rubber (Heffer and Sons, Cambridge).

Marinozzi, V. (1964), Cytochimie ultrastructurale du nucléole-RNA et protéines intra-nucléolaires, J. Ultrastruct. Res. *10*, 433.

Persson, A. (1972a), Equipment for cryo-ultramicrotomy. The LKB CryoKit, J. Microscopie *13*, 162.

Persson, K. G. A. (1972b), Cryo-ultramicrotomy instrumentation, Proc. 5th Eur. Reg. Conf. Electron Microscopy, Manchester, p. 254.

Pfeifer, U., E. Poehlmann and H. Witschel (1974), Kinetics of the accumulation of lead phosphate in acid phosphatase staining, Proc. 2nd Int. Symp. Electron Microscopy and Cytochemistry, Drienerloo, (North-Holland, Amsterdam) p. 25.

Sevéus, L. (1969), Cryo-ultramicrotomy, Process Biochem. *4*, 35.

Stumpf, W. E. and L. J. Roth (1965), Frozen sectioning below − 60 °C with a refrigerated microtome, Cryobiology *1*, 227.

Tokuyasu, K. T. (1973), A technique for ultracryotomy of cell suspensions and tissues, J. Cell Biol. *57*, 551.

Weavers, B. A. (1973), Combined transmission electron microscopy and X-ray micro-analysis of ultrathin frozen dried sections – an investigation to determine the normal elemental composition of mammalian tissue, J. Microscopy *97*, 331.

Yamataka, S. (1972), The cell (Saibou), *4*, 22. (In Japanese).

# Appendix

# List of suppliers

Commercial suppliers of equipment
and materials for sectioning

N.B.   The following list includes the names of suppliers known to the author at the present
time; she will be interested to hear of names and addresses of other suppliers for inclusion
in later editions.
  Suppliers are listed first from whom the majority of the smaller items may be obtained.
There follows a more comprehensive list with the items arranged in the same order as they
are mentioned in the text.
★ Denotes a manufacturer

## 1. General suppliers of materials and equipment for electron microscopy

(a) **Agar Aids for Electron Microscopy**
   (Agar Aids)
   (U.K. agents for Ladd)
   127a Rye Street
   Bishop's Stortford
   Hertfordshire
   England

(b) **Ernest F. Fullam, Inc.** (EFFA)
   P.O. Box 444
   Schenectady
   New York 12301
   U.S.A.

   **Graticules Ltd.**
   (U.K. agents for EFFA)
   Sovereign Way
   Tonbridge
   Kent
   England

**Touzart & Matignon**
(French agents for EFFA)
3 Rue Amyot
75 Paris 5
France

(c) **EMscope Laboratories** (EMscope)
   99 North Street
   London SW4 OHQ
   England

(d) **Ladd Research Industries Inc.** (Ladd)
   (U.S. agents for Agar Aids)
   P.O. Box 901
   Burlington
   Vermont 05401
   U.S.A.

(e) **Polaron Equipment Ltd.** (Polaron)
   (U.K. agents for Polysciences)
   4 Shakespeare Road
   Finchley
   London N3 1XH
   England

339

**Ted Pella Co.** (Pelco)
(U.S. agents for Polaron)
P.O. Box 510
Tustin
California 92680
U.S.A.

(f) **Polysciences Inc.** (Polysciences)
(U.S. agents for Polaron)
Paul Valley Industrial Park
Warrington
Pennsylvania 18976
U.S.A.

(g) **Taab Laboratories** (Taab)
52 Kidmore End Road
Emmer Green
Reading
Berkshire
England

**Extech International Corp.** (Extech)
(U.S. agents for Taab)
177 State Street
Boston
Massachusetts 02109
U.S.A.

(h) **E. M. Zairyo-Shya**
308 Yotsuya Sango Building
2 Saemoncho Shinjuku-ku
Tokyo
Japan

## 2. Ultramicrotomes (Chapter 2)

(a) *Cambridge Huxley Mark 2*

**Kent Cambridge Medical ★**
Histon Road
Cambridge CB4 3JA
England
(supplied only in U.K.)

(b) *LKB Ultrotome I, LKB Ultrotome III
and LKB Huxley*

**LKB-Produkter AB**
(Manufacturer of Ultrotomes I and
III)
S-161 25 Bromma
Sweden
(subsidiaries in Austria, Belgium,
Denmark, France, Italy and The
Netherlands)

*also*

**LKB Instrument GmbH**
D-8032 Lochham
AM Haag 11
Post Fach 121
West Germany

*also*

**LKB Instruments Ltd.**
232 Addington Road
South Croydon
Surrey CR2 8YD
England

*also*

**LKB Instruments Inc.**
12221 Parklawn Drive
Rockville
Maryland 20852
U.S.A.

(c) *Reichert OMU2 and Reichert OMU3*

**C. Reichert AG ★**
Hernalser Hauptstrasse 219
1170 Wien
Austria

**British American Optical Co. Ltd.**
Instrument Group
820 Yeovil Road
Slough
Buckinghamshire
England

*also*

EMscope                    (*see* 1c)

*also*

**American Optical Corporation**
Scientific Instrument Division
Buffalo
New York 14215
U.S.A.

(d) *Sorvall Porter-Blum MT1 and MT2*

**Du Pont Company** ★
Instrument Products Division
Sorvall Operations
Newton,
Connecticut 06470
U.S.A.

(e) *Tesla BS 490*

**Kovo** ★
Praha (Prague)
Czechoslovakia

**EXAL (Electronics and X-ray application Ltd.)**
Houndmills Estate
Basingstoke
Hampshire
U.K.

## 3. Vibration damping tables (§ 2.5 and § 6.1)

(a) *LKB Ultrostable*

**LKB** ★                  (*see* 2b)

(b) *Serva Bench Mark II*

**Barry Controls** ★
700 Pleasant Street
Watertown
Minnesota 02172
U.S.A.

(c) *Visolator*

**Meiritsu Seiki Co. Ltd.**
1535 Hirato Machi
Totsuka-ku Yokohama Shi
Japan

## 4. Equipment for making glass knives (Chapter 3)

(a) *LKB KnifeMaker 7800*

**LKB** ★                  (*see* 2b)

(b) *Sunkay Messer Knife Breaker*

**C. W. French Inc.**
58 Bittersweet Lane
Weston
Massachusetts 02193
U.S.A.

*also*

Polaron                    (*see* 1e)

*also*

**Sunkay Laboratories Inc.** ★
16-4, 2-chome Mita Shiba
Minatoku
Tokyo
Japan

(c) *Taab-Pyper Knife Maker*

Taab ★                     (*see* 1g)

(d) *Glass knife pliers*

EFFA                       (*see* 1b)
Kent Cambridge Medical     (*see* 2a)
Ladd                       (*see* 1d)

Polaron                    (*see* 1e)
Du Pont                    (*see* 2d)

(e) *Glass scoring wheel*

EFFA                       (*see* 1b)
Ladd                       (*see* 1d)
Polaron                    (*see* 1e)

(f) *Glass for making knives*

EFFA                       (*see* 1b)
Ladd                       (*see* 1d)
Taab                       (*see* 1g)
LKB                        (*see* 2b)
Du Pont                    (*see* 2d)

## 5. Troughs for glass knives (§ 3.7)

(a) *Metal Troughs*

EFFA                       (*see* 1b)
LKB                        (*see* 2b)

(b) *Tape for troughs*

EFFA                       (*see* 1b)
EMscope                    (*see* 1c)
Ladd                       (*see* 1d)
Polaron                    (*see* 1e)
LKB                        (*see* 2b)

3M Co.★
St. Paul
Minnisota
U.S.A.

(c) *Wax for sealing troughs*
   (dental wax)

EFFA                       (*see* 1b)
EMscope                    (*see* 1c)
Ladd                       (*see* 1d)
Pelco                      (*see* 1e)
Polaron                    (*see* 1e)
Taab                       (*see* 1g)
LKB                        (*see* 2b)

## 6. Diamond knives (§ 3.8)

**Friedrich Dehmer ★**
8202 Bad Aibling
Kolpingstrasse 8
West Germany

**E.I. Du Pont de Nemours
   & Co. (Inc.) ★**
Instruments Division
Wilmington
Delaware 19898
U.S.A.

*also*

**Du Pont de Nemours (France) S.A.**
9, Rue de Vienne
Paris 8e
France

*also*

**Du Pont Company (U.K.) Ltd.**
Wilbury House
Wilbury Way
Hitchin
Hertfordshire
England

**Akashi Seisakusho Ltd.**
CPO Box 1405
Tokyo
Japan

**GE-FE-RI**★
146/G Via Marittima 1
Frosinone
Italy

**IVIC**★
Apartitado 1827
Caracas
Venezuela

Ladd                                    (*see* 1d)

**Ernst Leitz**
Wetzlar
Germany

Polaron                                 (*see* 1e)

**Rawyler and Wolf**★
Pieterlen
Bern
Switzerland

**Rondikin Corp.**★
2003 Kalia Road
PH-7, Honolulu
Hawaii 96815
U.S.A.

## 7. Capsule Removers (§ 5.1)

(a) *BEEM capsule press*

**Better Equipment for Electron
Microscopy Inc. (BEEM)**★
P.O. Box 132
Jerome Avenue Station
Bronx
New York 10468
U.S.A.

EFFA                  (*see* 1b)
EMscope               (*see* 1c)
Polaron               (*see* 1e)
EM Zairyo-Shya        (*see* 1h)

(b) *Capsule splitter*

Polaron★              (*see* 1e)
Taab★                 (*see* 1g)

## 8. Razor blades (single-edged) (§ 5.2)

EFFA           (*see* 1b)
EMscope        (*see* 1c)
Pelco          (*see* 1e)
Polaron        (*see* 1e)
Polysciences   (*see* 1f)

## 9. Specimen block trimmers (§ 5.3)

(a) *Block shaper*

Kent Cambridge Medical ★   (*see* 2a)

(b) *Pyramitome*

LKB ★                      (*see* 2b)

(c) *TM60*

Reichert ★                 (*see* 2c)
British American Optical   (*see* 2c)
American Optical Corp.     (*see* 2c)

## 10.  Tweezers (§6.2 and §8.2)

General Suppliers              (*see* 1)

**Mason Morton Ltd.**
Fir Tree House
Headstone Drive
Wealdstone, Harrow
Middlesex HA3 5QS
England

## 11.  Grid storage boxes (§6.2 and §8.3)

(a) *BEEM*                                      (c) *Polaron*

| | | | |
|---|---|---|---|
| BEEM ★ | (*see* 7a) | Polaron ★ | (*see* 1e) |
| EFFA | (*see* 1b) | | |
| EMscope | (*see* 1c) | (d) *Metal type* | |
| Ladd | (*see* 1d) | | |
| EM Zairyo-Shya | (*see* 1h) | Ladd | (*see* 1d) |

(b) *LKB*

LKB ★                          (*see* 2b)

## 12.  Sectioning Aids (Chapters 6 and 7)

(a) *Pipettes and Syringes*

General Suppliers              (*see* 1)

(b) *Lens Tissues*

General Suppliers              (*see* 1)

(c) *Reichert Reflexomat*

C. Reichert                    (*see* 2c)

(d) *Westfall-Healy section mounter and diamond knife holder* (§6.5 and §8.3)

**Western Instrument Scientific Equipment ★**
RT 2 Box 220
Vacaville
California 95688
U.S.A.

(e) *Heat pen* (§7.7)

Polaron ★                      (*see* 1e)

## 13.  Grids (§8.1)

| *Supplier* | | *Type of Grid* | *Supplier* | | *Type of grid* |
|---|---|---|---|---|---|
| Agar Aids | (*see* 1a) | Athene | EMscope | (*see* 1c) | EFFA |
| EFFA | (*see* 1b) | EFFA ★ | | | Maxtaform |
| | | Maxtaform | | | Athene type |

| Supplier | | Type of grid | Supplier | | Type of grid |
|---|---|---|---|---|---|
| EM Zairo-Shya | (*see* 1h) | Maxtaform Veco | Pelco | (*see* 1e) | Pelco ★ |
| | | | Polaron | (*see* 1e) | Polaron ★ |
| | | | Polysciences | (*see* 1f) | Athene type |
| **Gilder** | | Gilder ★ | | | |
| Sinclair Gardens | | | **Smethurst-Highlight Ltd.** | | Athene ★ |
| West Kensington | | | Bolton | | |
| London W14 | | | Lancashire | | |
| England | | | England | | |
| Graticules | (*see* 1b) | Maxtaform ★ | Taab | (*see* 1g) | Maxtaform. |
| Ladd | (*see* 1d) | Athene type Micron | | | Micron |
| LKB | (*see* 2b) | Veco | **Veco Zeefplatenfabriek B.V.** | | Veco ★ |
| Mason & Morton | (*see* 10) | Mini-mesh ★ | Postbus 10 | | |
| | | | Karel Van Gelreweg 22 | | |
| **Micron Grids** | | | Eerbeek | | |
| Micron Grids | | Micron ★ | The Netherlands | | |
| 6 Boston Gardens | | | | | |
| Brentford | | | *and* | | |
| Middlesex | | | P.O. Box 2 | | |
| England | | | St. Neots PE19 3AZ. | | |
| | | | England | | |

## 14. Plastics for coating grids (§8.1)

(a) *Collodion (nitrocellulose)*

| | |
|---|---|
| EFFA | (*see* 1b) |
| EMscope | (*see* 1c) |
| Ladd | (*see* 1d) |
| Polaron | (*see* 1e) |
| Polysciences | (*see* 1f) |

**G. T. Gurr**
Searle Scientific Services
Coronation Road
Cressex Industrial Estate
High Wycombe
Buckinghamshire
England

**Fisons**
Scientific Apparatus
Bishop's Meadow Road
Loughborough
Leicestershire LE 11 ORG
England

(b) *Formvar (polyvinyl formal)*

| | |
|---|---|
| General Suppliers | (*see* 1) |
| Gurr | (*see* 14a) |

**Shawinigan Ltd.**
118 Southwark Street
London SE1
England

## 15. Coated grids (§8.1)

a) *Carbon-coated grids*

| | |
|---|---|
| Agar Aids | (*see* 1a) |
| EFFA | (*see* 1b) |
| Ladd | (*see* 1d) |
| Polaron | (*see* 1e) |

(b) *Formvar-coated grids*

| | |
|---|---|
| EFFA | (*see* 1b) |
| Ladd | (*see* 1d) |

## 16.  Attachments for cryo-ultramicrotomy (Chapter 12)

(a) *Cryokit* (for LKB Ultrotome I and III)

LKB ★                                    (*see* 2b)

(b) *FC 150 freezing attachment*

(for Reichert OMU3)

Reichert ★                               (*see* 2c)
British American Optical                 (*see* 2c)
American Optical                         (*see* 2c)

(c) *Frozen thin sectioner attachment*
    (for Sorvall Porter-Blum MT2B)

Du Pont ★                                (*see* 2d)

## 17.  Materials for cryo-ultramicrotomy (§ 12.3)

(a) *Bovine serum albumin* (BSA)

EFFA                                     (*see* 1b)
Gurr                                     (*see* 14a)
Polysciences                             (*see* 1f)
TAAB                                     (*see* 1g)

Research Products Division
**Miles Laboratories Inc**
Kanakee 111,
Illinois 60901
U.S.A.

(b) *Dimethyl sulphoxide* (DMSO)

EMscope                                  (*see* 1c)
Fisons                                   (*see* 14a)
Polaron                                  (*see* 1e)
Polysciences                             (*see* 1f)
Taab                                     (*see* 14g)

**Fisher Scientific Co. (Fisher)**
International Division
52 Fadem Road
Springfield
New Jersey 07081
U.S.A.

(c) *Ethylene glycol*

Fisons                                   (*see* 14a)
Fisher                                   (*see* 17b)

(d) *Gelatin*

EFFA                                     (*see* 1b)
Polysciences                             (*see* 1f)
Taab                                     (*see* 1g)
Fisher                                   (*see* 17b)

**SPO/S Rousselot 50**
Rue Boileau
Paris 16
France

(e) *FREON 12 (dichlorodofluoromethane)*
    *FREON 22 (monochlorodifluoro-*
    *methane)*
    *Sold as can-o-gas*

**Balzers AG**
FL 9496 Balzers
Liechtenstein

*also*

**Balzers High Vacuum Ltd.**
Northbridge Road
Berkhamstead
Hertfordshire
England

*also*

**Balzers High Vacuum Corporation**
P.O. Box 10816
Santa Ana
California 92711
U.S.A.

**Dean and Wood (London) Ltd.,**
83/85 Mansell Street
London E.1

*Sold as Halocarbon 12 and 22*

Air Products, speciality gases department
Stenor Street
Cobridge Burslem
Stoke-on-Trent

*FREON 22 sold as UCON*

**J. T. Baker Chemical Company**
Phillipsburg
New Jersey
U.S.A.

(*f*) *Arcton*

ICI Ltd.
Mond Division
Rocksavage Works
Runcorn
Cheshire
England

(g) *Isopantane*

BDH Chemicals Ltd.,
Poole
Dorset BH12 4 NN
England

(h) *Methyl Cellulose*

BDH                              (*see* 17g)

*Sold as Dow Methogel HG Powder*

**Dow Chemical**
International S.D.
Midland
Michigan
U.S.A.

(i) *Propane 7*

BDH                              (*see* 17g)

# Index for list of suppliers

Number referred to are the sections in the preceding list of addresses

# Subject index